Did Dinosaurs Survive?

Ancient Legends, Modern Encounters and the Search for

Living Dinosaurs

Table of Contents

By Andrei Nebeleac

www.nebeleac.com[1]

1. http://www.nebeleac.com

Copyright

Did Dinosaurs Survive?
Ancient Legends, Modern Encounters and the Search for Living Dinosaurs

This book investigates historical accounts, legends, eyewitness reports, cryptozoological claims, scientific hypotheses, and disputed evidence concerning the possible survival of non-avian dinosaurs and other prehistoric animals. The inclusion of a claim, encounter, legend, or theory does not mean that the author accepts it as scientifically established fact.

Where appropriate, conventional scientific explanations and alternative interpretations are presented alongside extraordinary claims. Readers should distinguish between verified scientific evidence, historical testimony, folklore, speculation, and unresolved reports.

Every reasonable effort has been made to present the material accurately. Scientific interpretations may change as new discoveries and research become available.

First Edition, 2026

Published independently.

Contents

Indigenous traditions, explorers' reports, giant reptiles, mysterious footprints, and the possibility of unknown animals surviving deep in the Amazon.

Chapter 8 — Papua New Guinea and the Search for Living Pterosaurs

The ropen, mysterious flying-creature reports, indigenous traditions, modern expeditions, and alternative explanations.

Chapter 9 — Modern Encounters With Supposed Dinosaurs

Reports from hunters, soldiers, fishermen, villagers, explorers and travelers who claimed to have encountered prehistoric-looking animals.

Chapter 10 — Photographs, Videos, Footprints and Alleged Evidence

The strongest claimed physical evidence, disputed images, tracks, bones, recordings, hoaxes, and modern methods of verification.

Chapter 11 — The Scientific Case Against Living Dinosaurs

Fossils, population genetics, breeding requirements, carcasses, DNA, environmental evidence, modern wildlife surveys, and the problems facing the survival hypothesis.

Chapter 12 — Somewhere, Could They Still Be Alive?

A final assessment of the evidence, the most plausible survival scenarios, Earth's remaining unexplored habitats, and what would constitute genuine proof.

Introduction — What If the Extinction Wasn't Total?

Sixty-six million years ago, an enormous catastrophe supposedly brought the age of the dinosaurs to an end. The fossil record strongly supports a mass extinction at the end of the Cretaceous. But does that necessarily prove that every population of every non-avian dinosaur disappeared at essentially the same geological moment?

This book investigates a much more controversial possibility: **could isolated populations of recognizable dinosaurs have survived much longer—and could any have survived into historical or even modern times?**

Rather than assuming the answer, we will investigate the evidence.

Ancient descriptions of dragons and enormous reptiles, indigenous traditions, unexplored forests, mysterious footprints, colonial reports, modern eyewitness encounters, cryptozoological expeditions, photographs, alleged physical evidence, and scientific objections will all be examined.

Chapter 1 — Did Every Dinosaur Really Die 66 Million Years Ago?

Sixty-six million years ago, something catastrophic happened to Earth. In geological terms, the change was abrupt. The great non-avian dinosaurs that had dominated terrestrial ecosystems for more than 150 million years disappeared from the fossil record. Ammonites vanished from the oceans. Numerous groups of plants and microorganisms suffered severe losses. By the time ecosystems recovered, the world had entered a new age.

The conventional scientific conclusion is clear: non-avian dinosaurs became extinct during the Cretaceous-Paleogene, or K–Pg, mass extinction approximately 66 million years ago.

But that statement contains an important complication.

Science can determine when dinosaur fossils disappear from the geological record. It cannot identify the moment when the **last individual dinosaur on Earth died**.

Those are not necessarily the same thing.

The Catastrophe

The strongest evidence connects the mass extinction with the impact of a large asteroid at Chicxulub, in what is now Mexico's Yucatán Peninsula.

The impact released extraordinary amounts of energy. Material blasted into the atmosphere spread around the planet. Dust, soot and aerosols reduced sunlight reaching Earth's surface. Temperatures and precipitation patterns changed. Photosynthesis was disrupted. Food chains began collapsing.

Large animals requiring substantial quantities of food were particularly vulnerable.

The geological boundary marking this catastrophe can be detected around the world. One of its most famous characteristics is an unusual concentration of iridium, an element relatively scarce in Earth's crust but more abundant in many meteorites.

Above this boundary, confirmed non-avian dinosaur fossils essentially disappear.

This is powerful evidence of extinction.

It is not, however, a census of every dinosaur alive at the time.

Finding the Last Dinosaur

Imagine that a dinosaur species contained 50,000 individuals scattered across a continent.

Only an extraordinarily small proportion of those animals would become fossils.

An animal normally dies and decomposes. Scavengers consume it. Bones are scattered or destroyed. Acidic soils dissolve them. Rivers transport them. Geological processes eventually erase enormous quantities of biological evidence.

Fossilization requires unusual circumstances.

Consequently, the fossil record is incomplete.

Suppose the youngest fossil of our hypothetical species dates to 66.2 million years ago. Does that prove the species became extinct 66.2 million years ago?

No.

It establishes that the species existed **at least until approximately that time**.

It could theoretically have survived longer without leaving fossils that have yet been discovered.

This problem is known as the **Signor-Lipps effect**.

The Signor-Lipps Effect

The Signor-Lipps effect describes a simple but important problem in studying extinction.

Because the fossil record is incomplete, the last fossil discovered from a species will normally occur before the actual death of the last member of that species.

Imagine gradually removing cards from a deck without knowing how many remain. Finding the last red card in your sample does not demonstrate that no red cards existed afterward. You simply haven't found them.

Around a mass-extinction boundary, this can create the appearance that species disappeared progressively even if many actually survived until much closer to the catastrophe.

The effect works in another interesting direction for our investigation.

If a few populations survived the initial disaster but became extremely rare afterward, their chances of entering the fossil record would become even smaller.

Therefore, paleontology faces two different questions:

Did non-avian dinosaurs suffer catastrophic extinction around 66 million years ago?

The evidence overwhelmingly says yes.

Did every surviving individual and every isolated population disappear immediately?

That is much harder to establish.

Survival Doesn't Mean Survival Until Today

This distinction is critical.

Suggesting that some dinosaurs might have survived the catastrophe for another thousand years is completely different from suggesting they survived another **66 million years**.

Consider a hypothetical valley containing a few hundred dinosaurs after the impact.

They survive.

The surrounding populations disappear.

For several generations, the isolated animals reproduce successfully. Eventually disease, food shortages, genetic problems or another environmental disturbance eliminate them.

Such a population might have survived the official extinction boundary without establishing a long geological dynasty.

This phenomenon is sometimes called a **dead clade walking**: a lineage survives the immediate extinction event but never truly recovers and disappears later.

Therefore, even if convincing evidence were discovered showing dinosaurs living thousands of years after the K–Pg boundary, it would not automatically support stories of dinosaurs surviving into human history.

But it would change an important detail of the extinction story.

Dinosaurs Above the Boundary

There have been controversial reports of non-avian dinosaur remains apparently found in rocks younger than the K–Pg boundary.

Such discoveries immediately attract attention because, if demonstrated convincingly, they could indicate that certain dinosaurs survived into the Paleocene.

But there is a major geological complication.

Bones can move.

Suppose a dinosaur bone was originally deposited 67 million years ago. Millions of years later, erosion exposes the fossil. A river carries it from its original sediment and deposits it inside younger material.

A future paleontologist discovers the bone surrounded by younger sediment.

At first glance, it appears to be a younger dinosaur.

But the dinosaur itself did not live when the surrounding sediment formed.

The fossil was **reworked**.

This is one reason alleged post-Cretaceous dinosaur fossils require exceptionally careful geological investigation.

Researchers must examine the surrounding sediments, preservation of the bone, chemical composition, erosion patterns and exact stratigraphic position.

An extraordinary discovery requires more than a dinosaur bone in apparently young rock.

Ideally, scientists would want multiple associated skeletons, footprints, eggs, nests or other evidence clearly formed after the extinction boundary.

So far, no generally accepted evidence demonstrates populations of non-avian dinosaurs continuing far into the Paleocene.

But the controversy illustrates how difficult determining an exact extinction date can be.

Could Some Have Survived Thousands of Years?

This possibility is considerably easier to imagine than survival for millions of years.

The asteroid did not necessarily kill every dinosaur directly.

Animals on the opposite side of Earth did not simply fall dead at the moment of impact.

The greater danger was what followed.

Darkness.

Cooling.

Wildfires.

Destroyed vegetation.

Disrupted rainfall.

Food shortages.

Collapsing ecosystems.

Some individuals almost certainly survived the initial hours, days and probably considerably longer.

The real question is whether any populations remained reproductively viable.

A surviving population would need males and females—or an appropriate reproductive structure—of breeding age. It would require sufficient food, nesting areas and tolerable environmental conditions.

Small omnivorous dinosaurs would theoretically have possessed advantages over gigantic specialized herbivores and predators.

An animal capable of eating insects, seeds, carrion and different vegetation would have more options than a giant herbivore dependent upon enormous quantities of living plant material.

Geography would matter as well.

Could particular islands, valleys, forests or other environments have experienced less severe ecological collapse?

Possibly.

But surviving the first catastrophe would only be the beginning.

The Population Problem

A species cannot survive indefinitely as a handful of individuals.

Imagine ten dinosaurs somehow surviving in an isolated valley.

Even if they reproduce, their descendants would face severe genetic problems. Inbreeding increases the probability that harmful recessive traits will become expressed.

Accidents become dangerous at the population level.

Five females dying during a drought could be catastrophic.

A disease outbreak could eliminate everyone.

A landslide could destroy a major nesting site.

An unusually cold season could finish what the asteroid started.

Larger populations are more resilient.

But larger populations also require more territory and resources—and would be more likely to leave fossils.

This creates a paradox for long-term dinosaur survival.

The population must be **small enough to escape detection but large enough to remain biologically viable**.

The longer survival continues, the harder that balance becomes.

A Million Years Later?

Now increase the hypothetical survival period.

Ten thousand years.

One hundred thousand.

One million.

Five million.

With every additional interval, the hypothesis becomes more demanding.

A dinosaur lineage surviving for a million years would represent thousands upon thousands of generations for many species.

It would spread, fluctuate and respond to environmental changes.

Individuals would die.

Rivers would bury carcasses.

Animals would leave footprints.

Eggs would be abandoned.

Teeth would be shed.

Populations might expand into new territories.

The probability that absolutely none of this would appear in the fossil record becomes progressively smaller.

That doesn't make survival mathematically impossible.

It makes it increasingly difficult to reconcile with available evidence.

And if dinosaurs survived not one million but **66 million years**, an even greater problem appears.

The world itself changed dramatically.

Surviving a Changing Planet

A dinosaur lineage continuing toward the modern era would have experienced enormous transformations.

Continents moved.

Mountain ranges rose.

Oceans changed.

Climates warmed and cooled.

Forests expanded and contracted.

Grasslands appeared.

Mammals diversified.

New predators evolved.

Ice ages came and went.

Eventually humans appeared.

A surviving dinosaur population would have needed to endure all of this.

It would also evolve.

This does not necessarily mean it would become an entirely different kind of creature. Evolution can sometimes preserve a recognizable body plan for extremely long periods. But a population reproducing for tens of millions of years would accumulate genetic and anatomical changes.

Therefore, a hypothetical dinosaur living today would not literally be genetically identical to its Cretaceous ancestors.

It might nevertheless remain recognizable as a dinosaur in the ordinary sense—just as a modern crocodilian is immediately recognizable as a crocodilian despite its lineage's enormous antiquity.

That possibility is the one most relevant to this book.

Could something recognizably descended from a non-avian dinosaur lineage have survived without undergoing the radical transformation associated with birds?

There is presently no convincing evidence that it did.

But we can define what would have been required.

The Perfect Refuge

For extremely long survival, dinosaurs would need refuges.

The ideal refuge would require a relatively stable climate, sufficient food and water, breeding territory, protection from major competitors and enough geographical isolation to prevent the population from being destroyed.

Later, it would also need protection from humans.

Large tropical forests seem like obvious candidates.

So do enormous wetlands.

Remote islands might provide isolation, although limited territory would restrict population size.

Mountain valleys could isolate animals but expose them to climatic changes.

Cave systems are frequently invoked in cryptozoological theories, although most dinosaurs were not adapted to permanent subterranean existence.

The refuge would need another remarkable characteristic:

It would have to remain connected enough to sustain a population while being isolated enough to prevent discovery.

For millions of years.

That is an extraordinary requirement.

The Fossil Record Is Incomplete—But Not Empty

Supporters of prehistoric-survivor theories sometimes make a valid observation and then take it too far.

The valid observation is:

The fossil record is incomplete.

Absolutely.

The unjustified conclusion is:

Therefore dinosaurs could have survived until today without leaving evidence.

The second statement does not automatically follow from the first.

Missing one thousand years is easier than missing one million.

Missing one isolated population is easier than missing a widespread species.

Missing a tiny animal is easier than missing a thirty-ton sauropod.

The size, ecology, population and proposed survival period all matter.

A scientifically serious investigation therefore cannot treat every survival scenario equally.

How Long Could They Have Lasted?

We can imagine a hierarchy of possibilities.

Dinosaurs surviving the immediate aftermath of the asteroid impact is virtually inevitable at the level of individual animals.

Populations surviving for years or generations is conceivable.

Survival for thousands of years becomes harder to demonstrate but cannot be dismissed merely because we have not found the final individual.

Survival for hundreds of thousands of years would require much stronger evidence.

Survival millions of years into the Paleogene would require extraordinary paleontological evidence that has not been established.

Survival through the entire Cenozoic and into human history would be an extraordinary claim requiring extraordinary physical proof.

And yet this is precisely where our investigation becomes stranger.

Because humans have been telling stories about enormous reptilian creatures for thousands of years.

They called them dragons.

Serpents.

Monsters.

Beasts of the forests.

Creatures of lakes and swamps.

In modern times, some witnesses have given these creatures another name:

dinosaurs.

The fossil record does not currently demonstrate that non-avian dinosaurs survived into the age of humans.

But if we want to investigate the possibility seriously, fossils are only the beginning.

The next question is even more difficult:

If dinosaurs survived, where could they possibly have hidden?

Chapter 2 — Could a Dinosaur Population Remain Hidden?

Finding one living non-avian dinosaur would be one of the greatest scientific discoveries in history, but from a biological perspective, one animal would raise an immediate problem: where did it come from? If dinosaurs had survived from the end of the Cretaceous to the present, they could not have done so as a succession of isolated individuals. Somewhere, generation after generation, breeding populations would have been required. For roughly 66 million years, animals would have needed to reproduce, find food, avoid disease and environmental catastrophes, and maintain sufficient genetic diversity to prevent the population from collapsing.

This requirement makes the living-dinosaur hypothesis considerably more difficult than simply asking whether one animal could hide in a remote forest.

The Minimum Population Problem

There is no universal number representing the minimum population necessary for a species to survive. It depends on reproductive rate, lifespan, genetic diversity, sex ratio, environmental stability and many other factors. A population of several dozen animals might persist temporarily under favorable circumstances, but remaining viable for thousands of generations is a very different challenge.

Very small populations face inbreeding. Closely related animals reproduce with one another, increasing the probability that harmful genetic variants will become expressed. Genetic diversity declines, reducing the population's ability to adapt to diseases or environmental changes. Random events become disproportionately dangerous. A flood that kills twenty animals means little to a population of 100,000 but could destroy a species containing only forty.

A dinosaur lineage surviving for tens of millions of years would therefore probably require considerably more than a few individuals hiding in a valley. It would need a functioning population, and perhaps multiple connected populations exchanging individuals.

That creates the central contradiction of the hypothesis: **the population must be large enough to survive but small or inaccessible enough to remain undiscovered.**

Food Leaves Evidence

Size would matter enormously. A hypothetical population of small carnivorous dinosaurs weighing several kilograms would be much easier to sustain than a herd of enormous sauropods.

A large herbivore requires substantial quantities of vegetation. Hundreds of such animals would alter their environment by creating trails, breaking vegetation, producing enormous quantities of dung and leaving footprints. Predators would create different evidence: kills, bones, territorial behavior and interactions with other wildlife.

Consider a sauropod-like creature weighing several tonnes. Even without knowing its precise metabolism, it would require a substantial and reliable food supply. A breeding population would need enough productive habitat to support adults and juveniles continuously. The animals would die eventually, leaving skeletons containing hundreds of large bones.

Smaller dinosaurs present a less extreme problem. An animal the size of a chicken, dog or small deer could occupy dense vegetation and leave relatively subtle traces. If any non-avian dinosaur lineage had survived undetected, a small forest species would therefore be substantially more plausible than a *Tyrannosaurus*-sized predator.

But where could even such a population hide?

The Congo Basin

Central Africa immediately attracts attention because it contains one of Earth's greatest tropical wildernesses. The Congo Basin covers an enormous region containing dense rainforest, rivers, wetlands and swamp forests. Some areas are difficult to traverse even today.

This environment is particularly relevant because Central Africa has produced famous traditions of mysterious large animals, most notably Mokele-mbembe, which some cryptozoologists have interpreted as a surviving sauropod-like dinosaur.

Dense tropical vegetation can hide surprisingly large animals from direct observation. A satellite image may show the forest canopy perfectly while revealing almost nothing about an animal walking beneath it. Researchers studying known species can spend months without seeing them directly, relying instead on tracks, dung, camera traps or environmental DNA.

Yet hiding from occasional human observation is not the same as remaining scientifically invisible for millions of years. A large breeding population should leave biological traces. The Congo remains one of the strongest geographical settings for living-dinosaur speculation, but geography alone is not evidence that dinosaurs inhabit it.

The Amazon

The Amazon rainforest presents a similar possibility on an even larger scale. Its forests, wetlands and immense river systems contain extraordinary biodiversity, and new species continue to be described.

Some regions are remote enough that scientific surveys remain incomplete. Dense vegetation restricts visibility to surprisingly short distances, while rivers provide enormous transportation corridors for aquatic and semiaquatic animals.

Could an unknown large terrestrial animal exist there?

Possibly. Unknown species unquestionably remain to be documented. The more difficult question is how large such an animal could become before leaving unmistakable evidence.

A small dinosaur-like animal living high in the canopy or deep within inaccessible forest would be far easier to overlook than a herd of multi-ton herbivores. Indigenous communities would also be expected to know about conspicuous large animals long before visiting scientists discovered them, making local zoological knowledge particularly important when evaluating alleged encounters.

New Guinea and Indonesia

New Guinea contains rugged mountains, deep forests and isolated ecosystems. New species have repeatedly been documented there, and some regions remain exceptionally difficult to survey thoroughly. Indonesia adds thousands of islands, creating natural laboratories of geographical isolation.

The discovery of the Komodo dragon illustrates an important point—not because it is a dinosaur, which it certainly is not, but because large reptiles can inhabit restricted island environments and remain poorly known to outsiders for long periods while being familiar to local inhabitants.

Islands, however, create another problem. A dinosaur population trapped on a small island would have limited territory and limited genetic exchange. Natural disasters, volcanic eruptions, climate changes and human colonization could eliminate it. Large islands such as New Guinea offer more possibilities, but once again a population surviving for tens of millions of years would represent an extraordinary biological continuity.

Mountains and Lost Valleys

Another favorite refuge is the isolated mountain valley. Mountain systems can produce ecological islands separated by cliffs, glaciers, forests or extreme elevations. Small populations can remain genetically separated for long periods, producing species found nowhere else.

The romantic image of a "lost world" therefore contains a small element of biological reality. Isolated ecosystems exist.

But a hidden valley capable of supporting a breeding population of giant dinosaurs for millions of years is much less plausible. Satellite mapping has eliminated the possibility of enormous completely unknown valleys. What remains possible are poorly surveyed habitats containing species that humans have not yet scientifically catalogued.

Again, scale becomes decisive.

Could Dinosaurs Live Underground?

Caves provide perhaps the most dramatic hiding place. Earth contains enormous cave systems, some extending hundreds of kilometers, and many subterranean passages remain unexplored.

Could dinosaurs have retreated underground?

For most large dinosaurs, this would create more problems than it solves. Deep caves generally contain limited food because sunlight cannot support ordinary plant growth. Large herbivorous dinosaurs could not survive permanently in such environments. Large predators would require substantial prey populations, creating another ecological requirement.

Smaller animals could use caves as shelters while feeding outside, however. A hypothetical surviving dinosaur would not necessarily need to spend its entire life underground. Caves could provide nesting areas, protection and daytime refuges, much as they do for many living species.

Satellites Have Mapped Earth—but Not Every Animal

The claim that "there are no unexplored places left" is misleading. Virtually every significant landmass has been mapped, but mapping geography is fundamentally different from cataloguing biology.

Satellites can identify mountains, rivers, forest cover and sometimes individual large animals in favorable conditions. They cannot routinely look through dense tropical canopy and identify every creature underneath it. Even field biologists frequently detect elusive animals indirectly.

Camera traps, drones, acoustic sensors, environmental DNA and increasingly sophisticated satellite technology are steadily shrinking these biological blind spots. This makes the continued concealment of large species progressively harder.

Nevertheless, science continues discovering animals previously unknown to formal zoology. Many are insects, fish, amphibians and small reptiles, but occasionally larger mammals and other conspicuous animals are scientifically described surprisingly late.

That fact establishes that **undiscovered large animals are possible**. It does not establish that surviving dinosaurs are among them.

The Size of the Mystery

Ultimately, the probability depends heavily on what kind of dinosaur we imagine.

A hidden population of gigantic sauropods would be extraordinarily difficult to reconcile with the absence of carcasses, bones, tracks, feeding damage and reliable photographs. A population of large predatory dinosaurs would face similar problems.

A small nocturnal dinosaur living in dense tropical forest presents a fundamentally different scenario. Hundreds or even thousands might occupy a relatively restricted region while rarely being seen directly. If their remains decomposed rapidly in acidic tropical environments, the fossil and skeletal evidence might also be poor.

Yet 66 million years remains an enormous obstacle. Such animals would have endured countless climatic changes, geological transformations, diseases, competitors and ecological disruptions.

Their descendants would also have evolved, even if they retained a recognizably dinosaurian body form.

The question therefore has no simple answer. Could Earth still contain a scientifically undocumented large animal? Almost certainly the possibility cannot be excluded. Could it contain a hidden population of recognizable non-avian dinosaurs descended from Cretaceous survivors? There is currently no verified evidence demonstrating this, and the biological requirements make the hypothesis increasingly difficult as animal size increases.

But if such creatures did survive, the search should concentrate not on imaginary blank continents but on Earth's remaining biological blind spots: **dense equatorial forests, enormous wetlands, isolated islands, inaccessible mountain ecosystems and regions where humans can map the landscape from above without knowing everything that moves beneath the trees.**

And those are precisely the places from which some of the world's strangest stories continue to emerge.

Chapter 3 — Dragons: Humanity's Memory of Dinosaurs?

Long before anyone used the word *dinosaur*, humans were telling stories about enormous reptilian creatures. They appeared in China, Europe, the Middle East, Africa, India and the Americas. Some possessed wings; others crawled like gigantic serpents. Some lived in caves, mountains or swamps, while others inhabited rivers or seas. They could be guardians, monsters, gods, symbols of royal authority or enemies that heroes were expected to destroy.

The similarities are intriguing. Dinosaurs were enormous reptiles in the broad popular sense, and ancient people imagined enormous reptile-like monsters. Could there be a connection?

The tempting explanation is that humanity somehow remembered dinosaurs. But humans appeared tens of millions of years after the accepted extinction of non-avian dinosaurs. If conventional paleontology is correct, nobody ever watched a *Tyrannosaurus* walk across a valley or encountered a living sauropod. Something else must explain the dragon traditions.

Unless, of course, some prehistoric reptiles survived much longer than currently demonstrated.

The Chinese Dragon

China possesses perhaps the world's most famous dragon tradition. Chinese dragons are usually elongated creatures with serpentine bodies, scales, claws and elaborate heads. Unlike the destructive dragons of many European stories, they were often associated with rain, rivers, fertility, imperial authority and good fortune.

Their anatomy does not correspond neatly to any known dinosaur. They frequently combine characteristics of several animals, suggesting that they are mythological composites rather than zoological descriptions.

China, however, is also extraordinarily rich in dinosaur fossils.

Long before paleontology developed, Chinese people encountered large fossilized bones. Some fossils were historically identified as "dragon bones" and were even collected for traditional medicinal purposes.

This provides one plausible mechanism for the development or reinforcement of dragon traditions. Imagine a farmer centuries ago discovering an enormous fossilized vertebra. Without geological knowledge, the obvious conclusion might be that it belonged to an enormous creature that once inhabited the region.

Finding fossils does not require understanding extinction.

The discovery could become a story, the story a legend and the legend eventually part of a much larger cultural tradition.

Europe's Dragons

European dragons frequently appear more physically aggressive. Medieval art and literature depict enormous serpents, winged reptiles and four-legged monsters. They inhabit caves, guard treasure, terrorize settlements and are killed by saints or heroes.

The familiar four-legged, two-winged dragon of later European art resembles no dinosaur. Dinosaurs did not possess an additional pair of limbs transformed into wings while retaining four walking legs.

But earlier European dragon traditions were far less standardized.

Some "dragons" were essentially enormous snakes. Others had two legs. Some were aquatic. The Germanic and Scandinavian **lindworm** was often portrayed as an elongated reptilian creature. British folklore contains numerous "worms"—a word historically connected with serpentine dragons.

This variety matters because the modern dragon image may obscure what earlier witnesses or storytellers actually meant when they used words later translated as *dragon*.

A medieval account saying that someone encountered a dragon does not automatically describe a giant winged monster breathing fire.

The original animal might have been described much more simply.

Draco: The Monster Behind a Word

The Greek *drakōn* and Latin *draco* were associated with serpents and dragon-like creatures. The meaning evolved considerably through centuries of mythology, religion, literature and artistic representation.

This creates a serious problem when investigating historical reports.

Words change.

A writer describing a *draco* might mean something resembling a snake, an enormous serpent, a supernatural monster or a symbolic creature. A modern reader may unconsciously imagine a medieval fantasy dragon even when the original author had something completely different in mind.

Therefore, historical dragon reports must be examined in their original linguistic and cultural context before being interpreted zoologically.

The Slavic Zmey and the Romanian Balaur

Eastern Europe developed its own rich traditions.

Slavic cultures contain stories of the **zmey**, **zmaj**, **żmij** and related dragon or serpent beings. Their characteristics vary enormously by region. Some have multiple heads, fly through the sky or possess supernatural powers. Others appear more like enormous serpents.

Romanian folklore contains the **balaur**, usually portrayed as a gigantic dragon or serpent, frequently with multiple heads.

Interestingly, modern paleontology has borrowed the name. A remarkable theropod discovered in Romania was named *Balaur bondoc*.

This does not demonstrate that Romanian folklore preserved memories of living dinosaurs. The scientific name was deliberately inspired by the mythological creature.

But the coincidence illustrates how naturally ancient monster traditions and dinosaur discoveries become associated in modern imagination.

Africa's Giant Reptiles

Africa presents a more complicated case because stories of mysterious large reptiles continue into relatively modern times.

Central African traditions include creatures associated with rivers, forests and swamps. The most famous is Mokele-mbembe, frequently described in modern cryptozoological literature as possessing a large body, long neck and long tail.

Other reported creatures have been compared with horned dinosaurs or animals possessing plates along their backs.

But extreme caution is necessary.

When Western explorers ask local inhabitants whether they have seen an animal resembling a dinosaur—and especially when witnesses are shown dinosaur illustrations—the investigation can become contaminated. People may interpret unfamiliar questions differently, identify an illustration approximately rather than precisely, or adapt traditional stories to the expectations of visitors.

Nevertheless, Central African reports deserve separate investigation precisely because some are presented not merely as ancient mythology but as descriptions of animals supposedly encountered within living memory.

We will return to them later.

Mesopotamian Monsters

The civilizations of ancient Mesopotamia also imagined formidable reptilian and serpentine beings.

One famous example is **Mušḫuššu**, a composite dragon-like creature associated particularly with Babylon. It possesses a long neck, scaled body, tail and unusual limbs combining characteristics of different animals.

To modern eyes, ancient representations can occasionally appear vaguely dinosaur-like.

But resemblance is dangerous evidence.

Humans can construct imaginary animals by combining known creatures. A serpent's body, lion's feet, eagle's claws and crocodile-like scales can produce something that accidentally resembles a prehistoric animal.

No surviving Mesopotamian description provides scientific evidence that people encountered non-avian dinosaurs.

Still, the repeated association between enormous serpentine creatures and supernatural power demonstrates how deeply the dragon concept penetrated early civilization.

India's Nāgas

Indian traditions contain **nāgas**, powerful serpent beings associated with water, underground realms, fertility and supernatural forces.

They are generally much closer to gigantic or supernatural snakes than to dinosaurs. In art, nāgas may appear as enormous cobras or beings combining human and serpent characteristics.

India contains large snakes and crocodilians, making it unnecessary to invoke prehistoric survivors to explain why serpent mythology developed there.

But nāga traditions contribute to the global pattern: again and again, civilizations imagined enormous reptiles occupying dangerous or inaccessible environments.

Water.

Caves.

Underground worlds.

Deep forests.

Places humans could enter but never completely control.

Feathered Serpents of Mesoamerica

Across the Atlantic, Mesoamerican civilizations developed another striking image: the feathered serpent.

Quetzalcoatl among the Aztecs and Kukulkan among the Maya are among the best-known manifestations of this tradition.

The combination of reptile and feathers inevitably attracts modern comparisons with feathered dinosaurs.

But similarity does not establish historical connection.

Feathers possessed enormous symbolic importance in Mesoamerican cultures, while serpents were familiar living animals. Combining them into a supernatural being required no knowledge of dinosaurs.

Nevertheless, the comparison demonstrates an important psychological phenomenon. Humans repeatedly created creatures by combining characteristics of animals representing different environments—the serpent of the ground and the bird of the sky.

Millions of years earlier, nature itself had produced feathered dinosaurs.

The resemblance is fascinating, but fascination is not proof.

Were Fossils Responsible?

Fossils provide one of the strongest natural explanations for some dragon traditions.

Ancient people encountered enormous bones.

They had no concept of dinosaurs, geological time or mass extinction. They attempted to explain what they found using the intellectual framework available to them.

A giant fossil bone meant a giant creature.

A large fossil skull might suggest a monster.

Tracks preserved in stone could appear to demonstrate that enormous creatures once walked nearby.

Over generations, real discoveries could merge with folklore.

This hypothesis does not require ancient humans to encounter living dinosaurs. It requires only that they encounter their remains.

Known Animals Could Create Monsters

Living animals provide another explanation.

Imagine seeing a crocodile for the first time without knowing crocodiles exist. A large individual emerging from muddy water could easily inspire stories of monsters.

Giant snakes provide an even more obvious source.

Monitor lizards can become large and intimidating. The Komodo dragon demonstrates that enormous predatory lizards are not purely imaginary. Whales, large fish and unusual marine animals could contribute to sea-dragon traditions.

Add exaggeration and oral transmission.

A three-meter snake becomes five meters.

Five meters becomes ten.

After generations, the creature acquires horns.

Then wings.

Then the ability to breathe fire.

A biological encounter has become mythology.

Could Memory Survive for Millions of Years?

There is one explanation that can almost certainly be rejected: humans directly remembering dinosaurs from 66 million years ago.

Human cultural memory cannot extend remotely that far. Our lineage did not coexist with the famous Cretaceous dinosaurs under the accepted geological chronology.

Therefore, if any dragon tradition originated from direct observation of a non-avian dinosaur, that dinosaur would itself have needed to survive into the human era.

That transforms the argument completely.

Dragon legends cannot prove dinosaur survival.

Instead, dinosaur survival would be required before certain dragon legends could potentially be interpreted as encounters.

The direction of evidence matters.

Myth, Memory or Animal?

Why, then, are dragons nearly universal?

There may be no single answer.

Humans everywhere encountered snakes, crocodilians, large predators, dangerous rivers and mysterious bones. We feared predators and imagined larger ones. Fossils demonstrated that gigantic creatures had once existed even when people did not understand what those remains represented. Stories traveled between cultures and changed with every retelling.

Some traditions developed independently; others influenced neighboring civilizations.

The similarities between dragons and dinosaurs therefore remain intriguing but inconclusive.

No dragon legend constitutes scientific evidence that dinosaurs survived into human history.

Yet legends should not automatically be dismissed as meaningless either. Folklore sometimes preserves memories of real places, natural disasters and animals, although those memories become transformed through generations.

The correct question is not:

Were dragons dinosaurs?

There is no evidence allowing us to say that they were.

The more useful question is:

Was every dragon entirely imaginary?

And if some stories began with encounters with real animals, what exactly did those people see?

To investigate that possibility, we must leave mythology behind and move toward reports involving claimed locations, witnesses and creatures supposedly encountered in the physical world.

That is where the search for living dinosaurs becomes considerably more controversial.

Chapter 4 — Ancient and Medieval Reports of Strange Beasts

Dragon mythology is one thing. Historical testimony is another. Across ancient and medieval literature we occasionally encounter writers who appear to describe enormous reptiles, serpents or unidentified beasts as real animals rather than supernatural symbols. These accounts are fascinating because some include locations, witnesses or physical details. Yet they present a major problem: ancient authors frequently mixed observation, hearsay, folklore and symbolism without making the distinctions expected from a modern zoologist.

If recognizable dinosaurs survived into human history, such reports might be among the first places to search. But each account must be examined according to what the original writer actually claimed, not according to what modern readers want the creature to have been.

Herodotus and the Flying Serpents

One of the most frequently discussed ancient accounts comes from the Greek historian Herodotus, writing during the fifth century BCE. While describing Egypt and Arabia, he mentioned "winged serpents." He claimed to have visited a place in Arabia near the Egyptian frontier where numerous bones belonging to these creatures could supposedly be seen.

Herodotus did not describe anything resembling a classic large dinosaur. His creatures were serpentine and apparently capable of flight. Later readers have occasionally compared them with pterosaurs, although pterosaurs were not dinosaurs.

Did Herodotus personally see living flying reptiles? His own account does not establish this. He reported seeing bones and repeated information about the animals. Some scholars have suggested that the story could have originated from misunderstood fossils, unusual snakes, bats or local traditions.

Nevertheless, the passage is interesting because Herodotus presented the creatures within a geographical and natural-historical discussion rather than simply as characters in a heroic myth.

Alexander's Campaigns and Giant Serpents

Stories surrounding Alexander the Great contain numerous strange animals. During campaigns toward India, Greek and later writers described enormous serpents supposedly inhabiting the region. Some versions became increasingly spectacular as Alexander's adventures were transformed into legend.

India does possess large snakes, including pythons, so there is no need to invoke surviving prehistoric animals. Ancient measurements were also notoriously unreliable, particularly when travelers repeated secondhand information.

This demonstrates one of the fundamental problems with historical cryptozoology. A real python eight meters long can become a fifteen-meter monster after several retellings. Centuries later, a modern reader encounters the final version and wonders whether the original witness saw an unknown species.

The earliest version is therefore always more valuable than the most dramatic one.

Pliny's Catalogue of Monsters

Roman naturalist Pliny the Elder attempted to describe the natural world in his *Natural History* during the first century CE. His enormous work contains genuine observations alongside extraordinary reports.

Pliny described giant serpents and unusual creatures from Africa and Asia. One famous tradition concerned an enormous serpent supposedly encountered by the Roman army during the First Punic War near the Bagradas River in North Africa. Later versions claimed that Roman soldiers required military weapons to kill it.

The story sounds astonishing, but its transmission is complicated. Ancient writers often repeated earlier authorities, and dramatic military stories could easily become exaggerated. Large pythons existed in Africa, while crocodiles could also inspire accounts of enormous river reptiles.

The case therefore demonstrates the difficulty of distinguishing an extraordinary animal from an extraordinary story.

The Roman Draco

Roman sources used the word *draco*, but translating it automatically as the modern English "dragon" creates confusion. The term could refer to a large serpent or serpent-like creature, and *draco* imagery also appeared on military standards.

A Roman author mentioning a *draco* was not necessarily claiming that a gigantic four-legged winged reptile existed.

Language evolved. Medieval writers inherited classical terminology and combined it with Christian symbolism and regional folklore. Over centuries, the dragon became increasingly elaborate.

Consequently, the word itself is weak evidence. Anatomy matters more.

Did the writer describe legs? Wings? Scales? Teeth? Size? Movement? Habitat? Behavior?

Those details are much more useful than the label attached to the creature.

Medieval Dragons Become Local

During the Middle Ages, European dragon stories often became remarkably specific geographically. Instead of monsters inhabiting distant India or Africa, dragons appeared near villages, caves, rivers and mountains.

England produced numerous traditions of "worms" and dragons. German-speaking Europe had lindworms. France had the *vouivre.* Scandinavia developed its own serpent and dragon traditions, while Eastern Europe possessed zmeys, zmajs and related creatures.

Many accounts followed a recognizable narrative: a monster appeared, livestock disappeared, villagers became terrified, and eventually a saint, knight or local hero killed the creature.

That structure should immediately make us cautious. It resembles folklore rather than zoological documentation.

But occasionally the supernatural elements are surprisingly limited.

Some creatures do not speak, cast spells or possess magical powers. They behave like animals. They occupy territories, eat livestock, hide in caves or wetlands and attack when approached.

These are the reports most interesting to an investigation of unknown animals.

The Lambton Worm

The Lambton Worm of northeastern England is a famous example. Later versions describe an enormous serpentine creature associated with the River Wear. It grows to gigantic size and eventually terrorizes the countryside before being killed.

The story is clearly surrounded by legendary material, and surviving versions were recorded long after the supposed events. There is therefore no reliable basis for treating the Worm as evidence of an unknown reptile.

Yet the creature's basic form—an enormous elongated animal associated with water—is much simpler than the conventional winged dragon.

Large fish, eels, snakes or exaggerated observations could have contributed to such traditions.

The same problem appears repeatedly: beneath an elaborate legend there may have been a real animal, but recovering that original event centuries later can be impossible.

The Bisterne Dragon

Another English tradition concerns the Bisterne Dragon in Hampshire. The creature supposedly lived around Burley Beacon and preyed upon animals before being killed by a local man.

Later accounts describe something recognizably dragon-like, but the historical evidence is weak. There is no preserved carcass, contemporary scientific description or reliably dated eyewitness testimony.

Could an unusual animal have inspired the story? Certainly.

Can we identify it as a dinosaur? No.

This distinction is essential throughout the investigation.

Central European Lindworms

Central European traditions may be more interesting anatomically. Lindworms were often depicted as elongated reptilian creatures with two limbs rather than the six-limbed structure of the classic European dragon.

Some historical illustrations look remarkably strange to modern eyes.

But illustrations are not photographs. Medieval and early modern artists routinely depicted known animals inaccurately. Crocodiles, elephants and whales could become almost unrecognizable when artists worked from verbal descriptions.

A bizarre-looking dragon illustration therefore tells us as much about artistic transmission as it does about the supposed animal.

Still, reports describing grounded, reptilian animals deserve greater attention than obviously supernatural creatures breathing fire and carrying humans through the sky.

Marco Polo's "Dragons"

Marco Polo provides one of the most frequently cited medieval examples. In describing parts of China, he reported enormous serpents with legs near the front of their bodies. According to versions of his account, the animals possessed large heads and formidable jaws and were hunted by local people.

Modern proponents of living dinosaurs sometimes present this passage as evidence that Marco Polo encountered dinosaurs.

That interpretation goes far beyond the evidence.

Large crocodilians provide an obvious comparison. Large monitor lizards offer another, although their geographical distribution and exact characteristics must be considered. Translation and manuscript variation further complicate the account.

Most importantly, Marco Polo's writings combine direct observation with information received from others. We cannot assume he personally examined every creature he described.

Nevertheless, his "serpents" are among the more interesting medieval reports because they are presented as ordinary dangerous animals rather than supernatural monsters.

Crocodiles: The Perfect Dragon Factory

Crocodilians probably explain many historical reptile-monster reports.

Imagine a medieval European traveler seeing a five-meter crocodile for the first time. It has armored scales, enormous jaws, claws and a powerful tail. It emerges unexpectedly from water and can kill a large mammal.

Describe that animal to someone hundreds of kilometers away.

Then allow the description to pass through several storytellers.

The result could easily become a dragon.

Crocodile specimens and skins also traveled through medieval and Renaissance trade networks, giving Europeans physical evidence that extraordinary reptiles genuinely existed somewhere beyond their familiar world.

Whales and Sea Dragons

Marine monsters present an even greater identification problem. Before modern marine biology, sailors occasionally encountered whales, giant sharks, enormous rays, oarfish and decomposing carcasses.

A partly decomposed whale can look astonishingly unlike a living whale. Skin disappears, tissues collapse and skeletal structures become exposed. Such carcasses have repeatedly generated sea-monster stories even in relatively modern times.

An oarfish several meters long can resemble a gigantic sea serpent. A whale glimpsed briefly between waves can appear to possess a completely different body shape.

Consequently, historical sea-dragon reports provide particularly weak evidence for prehistoric survival unless exceptionally detailed documentation exists.

The Reports That Remain Difficult

After removing supernatural stories, mistranslations, obvious folklore, crocodiles, snakes, whales and exaggerations, a small residue of historical reports can remain difficult to classify.

"Difficult to classify" does not mean "dinosaur."

That distinction cannot be emphasized enough.

Some descriptions are too incomplete to identify. Others survive only in manuscripts written decades or centuries after the alleged event. Some describe animals unlike known local species but could involve exotic animals transported through trade.

The most interesting reports share several characteristics: the creature behaves like an animal rather than a supernatural being; a recognizable location is provided; multiple people supposedly observe it; anatomical details are given; and the account appears relatively close in time to the alleged encounter.

Even then, without physical remains, identification remains speculative.

Witnesses or Storytellers?

Ultimately, ancient and medieval accounts must be divided into three categories.

Some authors described things they claimed to have personally observed.

Others recorded testimony received from travelers or local inhabitants.

A third group reproduced older stories, sometimes centuries after the supposed event.

These categories cannot be treated equally.

A soldier writing immediately after encountering an unidentified animal would be far more valuable than a monk recording a local dragon legend three centuries later.

Unfortunately, many spectacular cases belong to the second or third category.

This is why historical dragon reports have never overturned the paleontological consensus. None provides a securely dated skeleton, preserved tissue or other physical evidence demonstrating that non-avian dinosaurs survived into human history.

But the chronicles still contain a genuine mystery.

Ancient people encountered animals they did not understand. They found enormous bones. Travelers entered ecosystems containing creatures completely unfamiliar to them. Stories moved thousands of kilometers and survived for centuries.

Most strange beasts can probably be explained through this combination.

Perhaps all of them can.

Yet scattered among the obvious legends are descriptions of creatures presented not as demons or gods, but simply as **animals**—animals the writers believed genuinely existed.

Whether those reports represent crocodiles, giant snakes, lost species or something considerably stranger remains one of the most intriguing questions hidden inside the old chronicles.

Chapter 5 — Africa's Mokele-Mbembe

Deep in the Congo Basin, where immense rivers cut through rainforest and swamps stretch beyond easy human access, one of cryptozoology's most persistent creatures is said to live: Mokele-mbembe.

The name became internationally famous because Western explorers and writers increasingly described the animal as something resembling a surviving sauropod dinosaur: a massive body, long neck, long tail and an aquatic or semiaquatic lifestyle. It is often portrayed moving slowly through rivers and forest clearings, sometimes feeding on vegetation and sometimes reacting aggressively when approached.

If even a fraction of these reports described a real unknown animal, Mokele-mbembe would be extraordinary. If the sauropod interpretation were correct, it would overturn much of what we understand about dinosaur extinction.

But the evidence is far less straightforward than the legend.

The Congo as the Perfect Setting

The Congo Basin is one of the largest tropical forest regions on Earth. Dense vegetation, enormous wetlands, difficult transportation and limited scientific access to some areas have helped create the impression that almost anything might remain hidden there.

That impression is not entirely unreasonable.

Large mammals can be difficult to observe even when researchers know they are present. Forest elephants, gorillas and other animals may be detected more often through tracks, dung and local knowledge than through direct sightings.

This has encouraged a simple argument: if known animals can remain unseen for long periods, perhaps an unknown one could as well.

The problem is scale. A breeding population of very large animals should eventually leave substantial evidence. Trails, feeding damage, dung, carcasses, bones and repeated reliable observations would be expected.

That evidence has never been established for Mokele-mbembe.

Early Colonial Reports

Western interest developed from reports circulating during the colonial period. European travelers, administrators and hunters occasionally wrote about enormous unidentified animals supposedly inhabiting Central Africa.

Some stories referred to giant reptiles, strange tracks or mysterious beasts known to local communities.

These reports later became incorporated into the Mokele-mbembe narrative, but it is difficult to know whether they originally described the same creature.

One report might concern an elephant-like animal.

Another might describe a rhinoceros.

Another could involve a crocodile, hippopotamus or large snake.

Once cryptozoological writers began combining them, separate traditions could appear to form one continuous body of evidence.

This is one of the central problems surrounding Mokele-mbembe: **was there always one clearly defined animal, or did the modern creature emerge by combining multiple local stories?**

The Sauropod Image

The long-necked dinosaur interpretation became increasingly influential during the twentieth century.

Descriptions often sound remarkably familiar: a large rounded body, very long neck, small head and long tail.

That resembles the popular image of sauropods such as *Diplodocus* or *Apatosaurus*.

But this resemblance creates another problem.

Early twentieth-century reconstructions often portrayed sauropods as swamp-dwelling animals spending much of their time in water. Modern paleontology no longer considers this the standard interpretation. Sauropods were primarily terrestrial animals.

Yet Mokele-mbembe was frequently imagined according to the older swamp-sauropod model.

This raises an uncomfortable possibility: perhaps the description of the cryptid was influenced not by an actual surviving dinosaur, but by the way Western culture imagined dinosaurs at the time.

Expeditions Into the Swamps

Numerous expeditions have traveled into Central Africa searching for Mokele-mbembe.

Researchers, adventurers, missionaries and cryptozoologists interviewed local residents, traveled along rivers and searched remote wetlands.

Some expedition members returned convinced that something unusual existed.

They reported eyewitness testimony, unexplained noises, large tracks and accounts from hunters or fishermen who claimed to have seen strange animals.

But none returned with a specimen.

No skull.

No tooth.

No bone unquestionably belonging to an unknown living dinosaur.

No tissue.

No verified dung.

No clear sequence of photographs.

No body.

This absence does not automatically prove that every witness was mistaken. It does mean the case remains dependent primarily on testimony.

Eyewitness Accounts

Local eyewitness reports are the foundation of the mystery.

Descriptions vary.

Some witnesses describe an enormous animal with a long neck.

Others emphasize a long tail.

Some say it is reddish-brown or gray.

Some claim it lives mainly in rivers.

Others describe it entering the forest.

Some accounts say it is herbivorous but territorial and capable of attacking canoes.

If these descriptions represented independent observations of the same animal, they would be intriguing.

But independence is difficult to establish.

Once an idea becomes well known within a region, later witnesses may unconsciously describe what they already expect to exist.

Researchers themselves can also influence testimony.

This is particularly important when witnesses are shown pictures.

The Dinosaur Picture Problem

Some expeditions reportedly presented local inhabitants with illustrations of animals and asked which resembled Mokele-mbembe.

When witnesses selected sauropod images, investigators sometimes treated this as powerful evidence.

But such tests are methodologically weak.

Suppose a witness has described an unknown aquatic animal with a long neck. A researcher then presents drawings of an elephant, lion, rhinoceros, crocodile and sauropod.

The sauropod may simply be the closest visual option.

Selection does not prove exact identification.

There is an even greater problem if local people have already encountered previous explorers asking about dinosaurs. Over decades, the association between Mokele-mbembe and sauropods can enter local storytelling itself.

Eventually it becomes difficult to determine which came first: the traditional description or the Western dinosaur interpretation.

Translation and Meaning

Language adds another layer of uncertainty.

Names for animals do not always correspond neatly between cultures. A local term may describe a category of creatures rather than a single biological species.

The same word might refer to a dangerous river animal, a supernatural being or something encountered only in traditional stories.

Researchers unfamiliar with the language may misunderstand nuance.

Questions can also become leading without the investigator realizing it.

"Have you seen a strange animal in this river?" is very different from "Have you seen the long-necked animal that lives here?"

Translation through multiple intermediaries increases the risk further.

A witness speaks in a local language. A translator converts the statement into French. Another person records it in English. Years later a writer summarizes the account.

By then, subtle uncertainty can become apparent certainty.

Tracks and Footprints

Expeditions have occasionally reported large unexplained footprints or tracks.

Tracks can be valuable evidence, but only when documented properly.

A footprint should be photographed with scale references, mapped, cast if possible and examined within a complete trackway. Soil conditions should be recorded, and known animals should be excluded.

An isolated depression in swamp mud proves little.

Elephants, hippopotamuses and other large animals can leave distorted tracks. Water can enlarge impressions. Overlapping footprints may create entirely new shapes.

No Mokele-mbembe trackway has gained general zoological acceptance as evidence of an unknown giant reptile.

Photographs and Film

Photography should theoretically have transformed the case.

Instead, Mokele-mbembe photographs remain ambiguous.

Some alleged images show distant shapes on water. Others are unclear or poorly documented. None provides the anatomical detail required to identify a large unknown animal confidently.

This becomes more significant as camera technology improves.

Modern expeditions can carry high-resolution cameras, drones, camera traps and thermal equipment. If a large population regularly inhabits certain waterways, repeated photographic detection should eventually become possible.

Yet the decisive image has never appeared.

Strange Sounds and Recordings

Some investigators have reported unexplained calls or noises.

Tropical forests are acoustically complex environments. Birds, primates, frogs, mammals and insects produce sounds unfamiliar even to experienced visitors.

Recordings can therefore be interesting without being identifiable.

To demonstrate an unknown species, repeated recordings would need to be associated with direct observations, physical traces or other independent evidence.

An unusual sound by itself is not a dinosaur.

Could It Be a Known Animal?

Several familiar African animals could contribute to the reports.

Elephants frequently enter water, and only part of the body may remain visible. A raised trunk viewed from a distance can resemble a neck.

Hippopotamuses can produce enormous disturbances in rivers.

Crocodiles possess unmistakably reptilian bodies and can reach impressive sizes.

Large snakes could contribute to elongated-animal reports.

Floating logs can appear surprisingly animate when moved by currents.

Misidentification becomes particularly likely when observations occur briefly, through dense vegetation or at dusk.

Could It Be an Unknown Animal?

This possibility is more interesting than immediately jumping to dinosaurs.

The Congo Basin undoubtedly contains species that science has not yet documented. An unknown medium-sized mammal, reptile or aquatic animal is biologically possible.

A previously undescribed large species would be much more surprising, but not conceptually impossible.

This creates an important distinction.

Even if Mokele-mbembe were eventually proven to be real, it would not automatically be a dinosaur.

Science would first examine its anatomy, genetics and evolutionary relationships.

The discovery of an unknown large reptile would itself be spectacular without requiring a Cretaceous origin.

Could a Sauropod Survive There?

The sauropod hypothesis faces severe biological problems.

A population of multi-ton herbivores would require enormous amounts of vegetation.

They would create pathways.

They would damage plants.

They would leave dung in large quantities.

Individuals would die and leave enormous skeletons.

A viable population might contain hundreds or thousands of animals distributed across suitable habitat.

Hiding all of this for generations would be extremely difficult.

Furthermore, a lineage surviving 66 million years would not remain completely unchanged. Even if its descendants retained a sauropod-like form, evolutionary change would occur.

The hypothetical animal would therefore be better described as a surviving sauropod lineage than a perfectly preserved Cretaceous species.

Folklore or Cryptid?

There is another possibility: Mokele-mbembe may belong primarily to folklore.

Traditional societies everywhere tell stories about dangerous creatures inhabiting rivers, forests and forbidden places. Such stories can encode cultural rules, environmental knowledge and warnings about real dangers.

A dangerous river spirit might help keep children away from crocodile-infested water.

An enormous forest creature might symbolize a region considered dangerous or sacred.

Western visitors may then reinterpret these traditions literally.

A spiritual or mythological creature becomes an undiscovered animal.

An undiscovered animal becomes a dinosaur.

Over time, the transformation can become difficult to reverse.

The Evidence Today

After decades of expeditions and international attention, the situation remains remarkably unchanged.

People continue to report the creature.

The Congo remains vast.

Some regions remain difficult to survey.

But there is still no physical evidence accepted by zoology demonstrating the existence of Mokele-mbembe, much less identifying it as a surviving dinosaur.

The strongest evidence consists primarily of eyewitness testimony and local traditions.

The weakest part of the case is the absence of biological material.

For a creature claimed to represent a large breeding population, that absence is substantial.

What Would Prove It?

The mystery could be resolved quickly if the right evidence appeared.

A high-quality sequence of photographs showing anatomical detail.

Clear video from multiple angles.

Environmental DNA repeatedly detected and independently confirmed.

Fresh tissue.

Dung containing unknown DNA.

A recently dead specimen.

Bones with proteins or genetic material demonstrating modern age.

Ultimately, a living animal documented by multiple independent researchers.

Anything less would probably continue the debate rather than settle it.

Dinosaur, Unknown Animal or Story?

Four possibilities remain.

Mokele-mbembe could be a surviving dinosaur lineage.

It could be an unknown modern animal.

It could consist of misidentified known species.

Or it could be a cultural tradition gradually transformed into a zoological mystery by generations of explorers and writers.

The first possibility is unquestionably the most spectacular—and presently the least supported by physical evidence.

Yet Mokele-mbembe remains important because it represents the strongest popular version of the living-dinosaur idea. Unlike medieval dragons, it is supposedly encountered today, in a real ecosystem, by people who sometimes insist that the creature is not mythology.

That makes the Congo Basin one of the most obvious places where the hypothesis can be tested.

For more than a century, explorers have entered the forests hoping to return with proof.

So far, the animal has remained where it has always been: somewhere between **local tradition, eyewitness testimony and one of zoology's greatest hypothetical discoveries**.

Chapter 6 — Other African "Living Dinosaurs"

Mokele-mbembe dominates discussions about surviving dinosaurs in Africa, but it is only one member of a much larger collection of mysterious creatures reported from the Congo Basin and surrounding regions. Cryptozoological literature describes horned animals, creatures with plates along their backs and even enormous predators supposedly resembling theropod dinosaurs. The most frequently mentioned are Emela-ntouka, Mbielu-mbielu-mbielu, Ngoubou and the notorious Kasai rex. At first glance, the collection appears remarkable: if descriptions are taken literally, Central Africa seems to contain animals resembling several completely different dinosaur groups. Yet that apparent diversity is also a reason for caution. We must determine whether these creatures existed as independent traditions before outsiders began searching for dinosaurs, or whether Western interpretations gradually converted ambiguous local stories into a prehistoric bestiary.

Emela-ntouka — The Killer of Elephants

Emela-ntouka is usually associated with the Congo region and is commonly described as a large, dangerous animal possessing a prominent horn. Its name is often translated in cryptozoological accounts as something similar to "killer of elephants." Some reports portray it as roughly rhinoceros-sized, with a heavy body, powerful tail and a single horn projecting from its head. The animal supposedly inhabits swamps and waterways and is reputed to attack elephants or other large animals entering its territory.

The description inevitably encouraged comparisons with ceratopsian dinosaurs. A large animal with a horn immediately brings *Triceratops* and its relatives to mind, although the comparison is anatomically weak unless additional features are established. Ceratopsians possessed distinctive skulls, beaks and, in many species, elaborate frills. A witness simply reporting a large horned animal has not described a ceratopsian.

Known African wildlife provides alternatives. Rhinoceroses are obvious candidates, although their modern distribution does not neatly correspond with every alleged report. Hippopotamuses are extremely aggressive around water and could contribute to stories about powerful territorial animals. Elephants seen under poor conditions may also produce confusing descriptions. Another possibility is that Emela-ntouka represents a traditional creature whose characteristics became increasingly standardized through cryptozoological literature.

The phrase "killer of elephants" sounds impressive, but a dramatic name is not biological evidence. What matters is whether independent witnesses, questioned without leading suggestions, consistently describe the same anatomy. No verified skull, horn, footprint sequence, tissue sample or carcass has established Emela-ntouka as an unknown species.

Mbielu-Mbielu-Mbielu — Plates Above the Water

Even stranger is Mbielu-mbielu-mbielu, a creature associated in some accounts with the Likouala region of the Republic of the Congo. Descriptions collected by explorers have sometimes portrayed it as a large semiaquatic animal with structures resembling boards or plates projecting from its back.

For dinosaur enthusiasts the comparison was immediate: **Stegosaurus**.

The resemblance sounds astonishing. Stegosaurs possessed a spectacular series of plates along their backs, so a living animal displaying similar structures would naturally attract attention. Yet this case illustrates perfectly how easily interpretation can outrun testimony. Saying that something has visible structures along its back does not demonstrate that those structures are stegosaur plates. Crocodilian scutes, vegetation attached to an animal, partially visible objects in water or even misunderstanding between interviewer and witness could produce similar descriptions.

There is also an ecological problem. Stegosaurs disappeared long before the end-Cretaceous extinction itself, meaning that a literal surviving stegosaur would require a ghost lineage extending not merely 66 million years but well over 100 million years without a convincing fossil record documenting its continued existence. That makes the hypothesis even more extraordinary than survival of a Late Cretaceous dinosaur.

Reports of Mbielu-mbielu-mbielu remain interesting primarily because of the unusual dorsal structures attributed to it. But the evidence is almost entirely testimonial. There is no verified photograph showing a plated animal, no skeleton and no biological sample. Whether the original tradition actually described something resembling a stegosaur before outsiders introduced dinosaur imagery remains a crucial question.

Ngoubou — A Ceratopsian in Cameroon?

Ngoubou is another alleged African mystery animal that has been compared with horned dinosaurs. Reports associated particularly with Cameroon describe a large aggressive creature possessing horns, sometimes said to confront elephants. Certain versions describe more than one horn, encouraging comparisons with *Triceratops* or other ceratopsians.

Again, the problem is how the information was collected. When investigators already believe surviving dinosaurs may inhabit a region, interviews can unintentionally become searches for dinosaur characteristics. A witness mentions a horn; the investigator asks whether there were additional horns. A heavy body becomes "rhinoceros-like," then an illustration of *Triceratops* is introduced. Eventually the published story can sound considerably more dinosaur-like than the original testimony.

This does not require fraud. Human communication naturally reshapes uncertain information. Translation between languages increases the effect. If interviews occur decades after cryptozoological expeditions have already popularized dinosaur interpretations locally, genuinely independent testimony becomes increasingly difficult to obtain.

Nevertheless, reports of unfamiliar large animals in Central African forests cannot simply be declared impossible. The scientifically responsible position is narrower: an unidentified animal is conceivable, while identification as a ceratopsian requires evidence that currently does not exist.

Kasai Rex — Africa's Tyrannosaurus?

Kasai rex occupies a different category. Rather than a mysterious herbivore associated with local tradition, it is supposedly an enormous predatory reptile encountered in the Belgian Congo during the twentieth century. Popular versions tell of an explorer witnessing a gigantic dinosaur-like predator attacking or feeding upon a rhinoceros. Images allegedly connected with the incident have circulated widely, and the creature has been compared with *Tyrannosaurus* and other giant theropods.

It is a spectacular story—and one of the weakest cases under serious examination.

Accounts of Kasai rex are plagued by uncertain sourcing, inconsistent details and questionable imagery. Different versions provide different descriptions of the creature, and photographs circulated as evidence have been challenged as manipulated or constructed images. The supposed animal sometimes changes appearance between retellings. Reliable contemporary documentation capable of establishing the encounter is lacking.

There is also a geographical problem in some versions involving rhinoceroses in places where the details do not comfortably match known distributions. Such inconsistencies do not mathematically prove that no strange encounter occurred, but they seriously damage the credibility of the dramatic story presented today.

Kasai rex demonstrates an important principle for this entire investigation: **the more spectacular the claim, the more carefully its documentary history must be reconstructed**. A story repeated by hundreds of websites does not become hundreds of pieces of evidence. If every version ultimately traces back to one unreliable source, there is still only one source.

Are These Really Four Different Animals?

Taken together, these African cryptids create an extraordinary picture. Mokele-mbembe resembles a sauropod. Mbielu-mbielu-mbielu resembles a stegosaur. Ngoubou and Emela-ntouka resemble ceratopsians. Kasai rex resembles a giant theropod. It almost sounds as though a children's dinosaur encyclopedia has been transported into the Congo rainforest.

That coincidence should make investigators cautious.

If numerous dinosaur lineages had survived in Central Africa, they would require an entire hidden ecosystem containing breeding populations, prey species and enormous quantities of suitable habitat. The expected physical evidence would multiply accordingly. We should find tracks, bones, eggs, carcasses, dung and eventually DNA from several radically different prehistoric lineages.

Nothing approaching this has been verified.

A simpler possibility is that different African traditions concerning dangerous or mysterious animals were interpreted through Western dinosaur categories. A horned creature became *Triceratops*. A long-necked river creature became a sauropod. Structures along a back became *Stegosaurus* plates. A terrifying predator became *Tyrannosaurus*.

Once those associations appeared in books, documentaries and expeditions, they could influence subsequent reports.

The Importance of Independent Tradition

The strongest evidence would therefore come from traditions recorded **before** dinosaur comparisons were introduced. Researchers would need early manuscripts, colonial records, linguistic studies or independently collected oral histories demonstrating that local communities consistently described specific anatomical characteristics without being prompted by outsiders.

If several communities separated geographically and culturally described the same unusual animal independently, the case would become more interesting. If those descriptions matched an unknown biological creature rather than a supernatural being, stronger investigation would be justified.

Unfortunately, establishing such independence is difficult. Oral traditions change, colonial documentation is incomplete, translations can be unreliable and decades of cryptozoological interest have already influenced many locations.

What Remains?

Some cases weaken dramatically when their sources are examined. Kasai rex in particular offers little reliable evidence for a giant predatory dinosaur. Others, including Emela-ntouka, Ngoubou and Mbielu-mbielu-mbielu, remain harder to evaluate because they involve local testimony from regions where zoological documentation has historically been incomplete. But "unresolved" must never be confused with "confirmed prehistoric survivor."

Central Africa may still contain undocumented species. That is entirely plausible. Whether any resemble the creatures described in these traditions is unknown. Whether any represent dinosaur lineages surviving tens of millions of years is a far greater claim for which convincing physical evidence remains absent.

Yet the stories deserve investigation because they offer something more interesting than simple monster tales. They sit at the intersection of zoology, indigenous knowledge, translation, exploration and Western expectations. Somewhere within that mixture may be folklore, encounters with known animals, misunderstood descriptions, deliberate fabrication—and perhaps observations of species still awaiting scientific recognition.

The challenge is determining which is which.

If Africa truly contains an unknown large animal, the solution will not come from finding the story that sounds most like a dinosaur. It will come from **independent testimony, repeatable observations and physical evidence strong enough to survive scientific examination.**

Chapter 7 — The Amazon's Lost Monsters

If someone were asked to choose a place where a large unknown animal might remain hidden, the Amazon would be an obvious candidate. The basin covers millions of square kilometers across South America, containing immense forests, wetlands, tributaries and seasonally flooded landscapes. Humans have inhabited these environments for thousands of years, but scientific knowledge of Amazonian biodiversity remains incomplete. New species continue to be documented, particularly among insects, fish, amphibians, reptiles and small mammals. Against this background developed a much more extraordinary idea: perhaps the Amazon shelters not merely unknown species, but survivors from the prehistoric world.

Stories of enormous serpents, strange reptiles and unidentified creatures existed long before modern cryptozoology. Indigenous traditions, colonial accounts, missionaries, explorers and twentieth-century adventurers all contributed to a complicated collection of reports. As in Africa, however, the central challenge is separating independent observations from folklore, exaggeration, known animals and stories repeatedly transformed through retelling.

The Forest That Hides Everything

The Amazon's reputation as an unexplored wilderness can be exaggerated. Indigenous communities know their territories intimately, rivers have been traveled for centuries, and satellites now map the region continuously. Yet being geographically mapped does not mean being biologically catalogued. Dense canopy prevents direct observation of much of the forest floor, while seasonal flooding can make enormous regions difficult to traverse. Some animals are nocturnal, secretive or naturally rare.

A large creature could therefore remain unseen by visiting scientists for considerable periods. Remaining unknown to local inhabitants would be much harder. If a breeding population of large reptiles existed near rivers used for fishing and transportation, nearby communities would probably possess knowledge of them. Indigenous testimony is consequently important—but it must be recorded without forcing traditional creatures into Western categories such as "dinosaur."

The Giant Serpent Tradition

Perhaps the most persistent Amazonian monster is not dinosaur-shaped at all. It is the giant serpent.

Throughout South America there are stories of extraordinarily large snakes associated with rivers and deep pools. In Brazil, one famous tradition concerns the **Cobra Grande**, the Great Snake, sometimes described as reaching impossible dimensions and possessing supernatural characteristics. Related traditions appear among different communities under different names.

These accounts exist in a region where genuinely enormous snakes live.

The green anaconda is among the world's heaviest snakes and can reach impressive lengths. A large individual partly submerged in muddy water can be extremely difficult to estimate accurately. If only sections of its body appear between vegetation and water, witnesses may substantially overestimate its total length.

This creates a natural foundation upon which exaggeration can grow. A six-meter snake becomes eight meters, then twelve, then twenty.

After generations, an unusually large anaconda can become a monster capable of overturning boats.

But does every giant-serpent report reduce to anaconda exaggeration? That is harder to establish. Claims of extraordinarily large snakes continue to appear, although reliable measurements and physical specimens have repeatedly failed to support the most extreme sizes.

Percy Fawcett and the Giant Anaconda

One of the most famous Amazon explorers associated with extraordinary animals was British explorer Percy Harrison Fawcett. During his expeditions in South America in the early twentieth century, Fawcett reported encountering an enormous anaconda whose length he estimated at roughly nineteen meters.

The claim became famous and remains controversial.

No specimen was preserved. Estimating the length of a snake from a distance is difficult, and Fawcett's figure greatly exceeds reliably documented dimensions for living anacondas. Nevertheless, he was an experienced explorer rather than an anonymous storyteller, giving the account continuing appeal.

Fawcett also believed South America's interior still contained major geographical, archaeological and zoological discoveries. His eventual disappearance in 1925 while searching for a lost civilization transformed him into part of Amazonian legend himself.

His giant-snake story demonstrates the recurring problem: an apparently credible witness can report something extraordinary, but without a specimen or reliable measurement, testimony alone cannot determine whether the animal was genuinely exceptional or simply misjudged.

Colonel Fawcett's Other Strange Animals

Fawcett described other unusual creatures during his travels, including animals he believed might represent species unknown to science. Some were less spectacular than giant snakes, but collectively they reinforced the image of the Amazon as a place capable of producing zoological surprises.

This was not an absurd belief. Scientific expeditions have indeed documented previously unknown South American species.

The leap occurs when "unknown species" becomes "surviving dinosaur."

Nothing in the existence of undiscovered monkeys, rodents, fish or reptiles automatically makes a Cretaceous survivor more likely. A dinosaur lineage would carry an evolutionary history extending across 66 million years for which convincing post-Cretaceous evidence is absent.

Conan Doyle and *The Lost World*

The idea of prehistoric animals surviving on an isolated South American plateau entered popular culture spectacularly through Arthur Conan Doyle's 1912 novel *The Lost World*. Doyle imagined an expedition discovering dinosaurs and other prehistoric creatures surviving on an inaccessible plateau in the Amazonian region.

The story was fiction, but its setting reflected contemporary fascination with unexplored South America. Reports from explorers such as Fawcett helped create an atmosphere in which lost ecosystems seemed conceivable.

The novel subsequently influenced how people interpreted mysterious South American animals. Once the public possessed the image of a hidden plateau filled with dinosaurs, any strange footprint or unidentified jungle creature could be fitted into that framework.

Fiction did not create every Amazonian monster tradition, but it changed the vocabulary through which outsiders understood them.

Strange Footprints

Reports of enormous or unusual tracks occasionally appear in Amazonian exploration stories. Footprints are potentially valuable because they provide physical evidence even when the animal itself remains unseen.

Unfortunately, jungle tracks are difficult to interpret.

Mud deforms rapidly. Rain enlarges impressions. Several overlapping tracks can create one enormous shape. A tapir footprint can appear surprisingly strange to someone unfamiliar with it, while jaguars, peccaries, capybaras and other mammals leave tracks that deteriorate quickly in wet environments.

A genuine dinosaur trackway made by a living animal would be revolutionary. Investigators would need multiple consecutive footprints showing consistent anatomy, photographed with scales and preferably preserved as casts or three-dimensional scans.

No verified modern Amazonian trackway has met that standard.

Monsters in the Rivers

The Amazon River system contains enough genuine giants to produce monsters without requiring prehistoric survivors.

Black caimans can reach impressive dimensions and become formidable predators. Anacondas occupy aquatic environments.

Arapaima are enormous freshwater fish. Giant river otters can appear surprisingly unfamiliar when glimpsed briefly. Manatees possess large bodies that may be partially visible near the surface.

Floating vegetation, logs and wakes further complicate observations.

Rivers are particularly effective at hiding an animal's true anatomy because witnesses usually see only a fraction of the body. A head appears, disappears, and surfaces farther away. The observer mentally connects those positions into a creature potentially much longer than the actual animal.

This mechanism probably explains some sea-serpent and lake-monster reports around the world.

The Tapir Problem

The South American tapir illustrates how an unfamiliar known animal can generate extraordinary descriptions. It possesses a heavy body and a flexible elongated nose. Seen briefly at night, especially by someone unfamiliar with Amazonian fauna, it might be described inaccurately.

Tapirs also demonstrate another important point: large forest mammals can be remarkably difficult to observe directly. Researchers may know they inhabit an area primarily through camera traps, tracks and dung.

If an unknown dinosaur-sized animal occupied similar habitat, however, the problem changes with scale. A small animal could remain elusive. A herd of multi-ton creatures would leave much more substantial ecological evidence.

Sounds Without Animals

The Amazon at night can sound like another planet. Howler monkeys produce calls that travel enormous distances. Frogs, birds, insects and mammals create noises whose sources may be impossible for visitors to identify.

An explorer hearing a deep roar from dense forest may imagine a huge predator without ever seeing the animal responsible.

Sound is therefore among the weakest forms of cryptozoological evidence unless it can be recorded repeatedly and connected with physical observations. Unknown noises may indicate an unknown species, but they cannot establish its size, appearance or evolutionary ancestry.

Could Dinosaurs Actually Hide Here?

Among Earth's major ecosystems, the Amazon provides some of the best conditions for concealing an unknown terrestrial animal: enormous habitat, dense vegetation, abundant water, high biological productivity and regions difficult to survey continuously.

A small nocturnal reptile or dinosaur-like creature could theoretically be overlooked for a considerable time.

A sauropod-sized animal is another matter.

Large herbivores would consume substantial vegetation and create trails. Predators would require prey. Both would leave dung, tracks, carcasses and DNA. A viable breeding population would multiply these traces.

The problem is therefore not whether **one dinosaur could hide behind the trees**. It is whether an entire population could remain biologically invisible for countless generations.

That remains extremely difficult to reconcile with the evidence.

What Might Really Be There?

The most reasonable expectation is that the Amazon still contains many undescribed species, some perhaps surprisingly large. An unknown snake somewhat larger than reliably documented specimens is far easier to imagine than a surviving giant theropod. An unusual aquatic reptile is more plausible than a hidden sauropod population.

But science cannot identify an animal before evidence exists.

The correct position is neither "anything could be hiding there" nor "everything has already been discovered."

The Amazon remains incompletely known, but incompleteness is not evidence for dinosaurs.

Its monster stories probably have multiple origins: enormous known animals, misidentification, exaggeration, folklore, unusual individuals and occasionally species science has not yet catalogued.

Whether something more extraordinary lies among them remains unanswered.

The rainforest gives such possibilities unusual longevity because it can conceal an animal only meters away from an observer. Yet every year camera traps, drones, genetic sampling and expanding biological surveys make the hiding places smaller.

If a recognizable non-avian dinosaur lineage survives anywhere in South America, the Amazon would seem one of its best possible refuges.

But eventually even the deepest forest must surrender evidence.

And so far, the Amazon has produced extraordinary stories, enormous snakes and genuine zoological discoveries—but **no scientifically verified living dinosaur**.

Chapter 8 — Papua New Guinea and the Search for Living Pterosaurs

Papua New Guinea seems almost designed for stories about prehistoric survivors. Mountain ranges divide the island into isolated regions, tropical forests cover vast areas, caves penetrate limestone landscapes, and thousands of islands surround the main landmass. New species continue to be documented, while many communities possess rich traditions concerning unusual animals and supernatural beings. Among these stories, one creature has attracted exceptional attention from cryptozoologists: the **ropen**, an alleged large flying animal sometimes interpreted as a surviving pterosaur.

Pterosaurs were not dinosaurs. They belonged to a separate lineage of flying reptiles that appeared more than 200 million years ago and included species ranging from small animals to giants with wingspans comparable to small aircraft. The last known pterosaurs disappear from the fossil record around the end-Cretaceous mass extinction approximately 66 million years ago. If a pterosaur lineage survived into modern Papua New Guinea, it would represent one of the most extraordinary zoological discoveries imaginable.

Yet, as with living-dinosaur claims, the difference between an extraordinary story and extraordinary evidence is enormous.

What Is the Ropen?

Descriptions of the ropen are far from uniform. In modern cryptozoological accounts it is often portrayed as a large nocturnal flying creature with leathery wings, a long tail and sometimes a pointed structure or flange at the end of that tail. Some witnesses describe a long beak or elongated head. Others concentrate less on anatomy and more on the creature's enormous apparent wingspan.

The strangest characteristic is light.

Some accounts claim that the ropen produces or carries a bright glow while flying at night. Witnesses have described lights moving through the sky, sometimes disappearing over forests or the ocean.

This creates an immediate problem. Known pterosaurs are not believed to have been bioluminescent. If the mysterious lights are genuinely connected with an animal, a surviving pterosaur interpretation would require a biological ability for which there is no direct fossil evidence.

Alternatively, the light and the flying creature may represent two different phenomena that became combined into one legend.

Indigenous Traditions

Papua New Guinea contains hundreds of cultural and linguistic groups, so speaking of a single "Papua New Guinean tradition" is misleading. Different communities possess their own names and stories for unusual creatures.

Cryptozoological investigators have collected accounts of large flying animals from local residents, sometimes associating names such as ropen, duah and other regional terms with the same hypothetical species.

But this classification is controversial.

A local word may describe a large bird, bat, spirit, nocturnal creature or category of flying animals rather than one zoological species. Translation can transform an ambiguous description into something much more specific.

The crucial question is whether traditional descriptions independently contained pterosaur-like characteristics **before investigators introduced pictures of pterosaurs**.

If several isolated communities had independently described an animal with the same unusual anatomy—long tail, membranous wings, distinctive head and consistent behavior—the case would become significantly more interesting.

If descriptions became increasingly pterosaur-like only after Western expeditions arrived, the evidential value would be much weaker.

Modern Eyewitnesses

Reports are not limited to ancient traditions. Local inhabitants, missionaries, travelers and investigators have claimed sightings during the twentieth and twenty-first centuries.

Some describe enormous winged silhouettes crossing the sky. Others report creatures flying over water or disappearing into forested areas. Occasionally witnesses estimate wingspans far beyond those of ordinary local birds or bats.

Size estimates, however, are notoriously unreliable when an object is seen against an empty sky.

Without a nearby reference point, the human brain has difficulty determining both distance and size. A bird with a one-meter wingspan flying relatively close can appear similar to a much larger creature flying farther away.

Night observations are even less reliable. A silhouette seen for several seconds can acquire anatomical details through memory and expectation.

This does not mean witnesses are lying. A person can accurately report that they saw something unusual while being mistaken about what the object actually was.

The Pterosaur Comparison

Certain reported characteristics are undeniably suggestive. Membranous wings, elongated heads and long tails immediately resemble popular reconstructions of pterosaurs.

However, pterosaurs were extremely diverse. Some had long tails, while advanced groups generally possessed short tails. Some had spectacular cranial crests. Others did not. Their sizes and feeding strategies varied enormously.

A modern witness describing a generic "pterodactyl" may therefore be influenced more by films and illustrations than by actual paleontology.

The question should not be whether the creature resembles the popular image of a pterosaur. Investigators should ask whether its detailed anatomy matches a plausible pterosaur lineage.

That level of evidence has never been obtained.

Expeditions Searching for the Ropen

Several expeditions have traveled to Papua New Guinea specifically searching for living pterosaurs. Investigators have interviewed witnesses, watched the night sky, explored areas associated with sightings and attempted to photograph mysterious lights.

Participants have sometimes returned convinced that an unknown flying creature exists.

They have reported unusual lights, distant silhouettes and compelling testimony.

But no expedition has produced a specimen or clear anatomical documentation demonstrating a pterosaur.

This creates the same problem encountered with Mokele-mbembe. Repeated expeditions increase the number of stories without necessarily increasing the quality of evidence.

Ten unclear sightings are not equivalent to one clear specimen.

The Mysterious Lights

The luminous component of the ropen story deserves separate treatment because it is both fascinating and problematic.

Lights moving over forests or coastlines can have numerous causes. Aircraft, boats, distant human activity, meteors, planets near the horizon, atmospheric effects and other sources can appear mysterious under unfamiliar conditions.

Bioluminescent organisms provide another possibility, although explaining a large airborne light with them would require specific evidence.

Observers may also combine events. A strange light is seen one night; a large flying animal is observed on another occasion. Over time, the two become characteristics of the same creature.

To establish that the ropen itself produces light, investigators would need simultaneous observation of recognizable animal anatomy and luminescence.

That evidence does not presently exist.

Giant Bats

Papua New Guinea is home to large fruit bats, commonly called flying foxes. Some possess impressive wingspans and are frequently active around dusk or at night.

A large flying fox viewed under poor lighting can look remarkably unfamiliar. Its membranous wings immediately provide a superficial resemblance to a pterosaur.

Bats therefore offer one of the strongest conventional explanations for at least some ropen sightings.

Yet bats do not possess long reptilian tails or pterosaur-like heads. Reports containing those details cannot be explained simply by saying "flying fox" unless those characteristics arose through misperception or later memory.

Large Birds

Large birds provide another explanation. Papua New Guinea and neighboring regions support numerous substantial bird species. Frigatebirds, herons, eagles and other birds can produce unusual silhouettes depending on viewing angle.

A bird's legs trailing behind its body can sometimes appear like a tail. A long beak can create a prehistoric-looking profile. Distance can dramatically inflate estimated size.

Again, this could explain many reports without necessarily explaining every description.

Could a Pterosaur Population Survive?

Suppose, for argument's sake, that a pterosaur lineage survived the end-Cretaceous extinction.

One individual could not persist for 66 million years. Thousands upon thousands of generations would be required. A breeding population would need food, nesting locations and sufficient genetic diversity.

Flying animals have one disadvantage when attempting to remain hidden: they enter open sky.

A terrestrial creature can disappear beneath rainforest canopy. A large flying animal must periodically expose itself.

Papua New Guinea is remote, but it is inhabited. Aircraft cross it. Villages occupy coastlines and valleys. Cameras are widespread. A population of enormous flying reptiles should eventually produce repeated high-quality observations.

Dead individuals should also occasionally be found.

No verified modern pterosaur bones, eggs, tissue or DNA have emerged.

A 66-Million-Year Ghost Lineage

The fossil problem is equally formidable.

For living pterosaurs to exist today, their lineage would need to extend from the end of the Cretaceous through the Paleocene, Eocene, Oligocene, Miocene, Pliocene and Pleistocene into the present without producing a generally accepted fossil record.

That represents approximately 66 million years of missing evolutionary history.

Fossil records are incomplete, but such an enormous ghost lineage requires extraordinary evidence before it can be accepted.

A photograph of an ambiguous silhouette is nowhere near sufficient.

What Would Convincing Evidence Look Like?

A genuine living pterosaur would not be impossible to demonstrate scientifically.

High-resolution photographs from multiple angles could reveal anatomy. Repeated drone footage could document flight mechanics. Environmental DNA might detect an unknown vertebrate lineage. Feathers would argue strongly against a pterosaur identification, while membranous wing tissue would point elsewhere.

A carcass would settle the question almost immediately.

Researchers could examine bones, muscles, wing membranes and DNA, determining whether the creature represented a pterosaur lineage or something completely unexpected.

Nothing comparable has yet been produced.

Something in the Sky

The ropen mystery therefore occupies an uncomfortable middle ground. Papua New Guinea unquestionably contains large bats and birds capable of generating unusual sightings. Cultural traditions can change through translation, and mysterious nocturnal lights have many possible explanations. The lack of physical evidence makes a surviving pterosaur extraordinarily difficult to defend scientifically.

Yet it would also be unreasonable to conclude that every witness deliberately invented an encounter. Many probably saw something.

The real mystery is determining what.

Perhaps different witnesses saw different phenomena: flying foxes, birds, lights, unfamiliar aircraft or ordinary animals under extraordinary viewing conditions. Those separate observations may gradually have merged into the modern image of the ropen.

Perhaps an unknown flying species contributes to a minority of reports.

The most extraordinary possibility—that a lineage of pterosaurs survived for 66 million years and still flies above Papua New Guinea—remains unsupported by verified physical evidence.

But the story has an unusual advantage over tales of creatures hidden permanently beneath water or dense forest.

If the ropen exists, eventually it must enter the open sky.

And when it does, the question is whether the next witness will return with another story—or with an image clear enough to transform a legend into a biological discovery.

Chapter 9 — Modern Encounters With Supposed Dinosaurs

Ancient dragon stories can be dismissed as mythology and medieval accounts can be blamed on poor zoological knowledge, but modern reports create a more difficult question. During the twentieth and twenty-first centuries, hunters, soldiers, fishermen, pilots, missionaries, villagers, explorers and tourists have occasionally claimed to see animals that they described as prehistoric or dinosaur-like. These encounters supposedly occurred in Central Africa, South America, Papua New Guinea, Australia and even heavily populated countries.

Modern witnesses know what dinosaurs are. They also live in an age of cameras, scientific fieldwork and rapid communication. This should make modern reports easier to investigate. Paradoxically, it creates a new problem: because people already know what dinosaurs are supposed to look like, their memories and descriptions can be influenced by books, museums, documentaries and films. A witness saying, "It looked exactly like a *Tyrannosaurus*," therefore provides a recognizable description, but not necessarily an accurate identification.

The strongest cases must be approached almost like criminal investigations: locate the earliest account, establish what the witness originally said, identify independent observers, reconstruct the environment and determine whether physical evidence exists.

The Kasai Rex Encounter

One of the most spectacular alleged modern dinosaur encounters is Kasai rex, supposedly seen in the Belgian Congo during the 1930s. Popular versions describe a gigantic bipedal predator attacking or feeding on a rhinoceros. The creature is often compared with

Tyrannosaurus or another giant theropod, and supposed photographs have circulated online.

If genuine, the case would be revolutionary.

Unfortunately, Kasai rex begins to collapse when examined historically. Different retellings contain inconsistent details, the documentary trail is weak, and the famous images do not provide reliable evidence of an unknown animal. Even the circumstances of the encounter vary between versions.

This illustrates the first rule of investigating cryptozoological encounters: **a story repeated thousands of times remains one story if every version originates from the same questionable source.**

Internet repetition can create an illusion of independent confirmation.

Mokele-Mbembe Witnesses

Central Africa provides considerably more testimony than Kasai rex. Over decades, inhabitants of the Congo Basin have described a mysterious creature associated with rivers and swamps, later identified by some investigators as Mokele-mbembe.

Certain witnesses have described a long neck, large body and tail. Some expedition members have claimed brief observations themselves.

The number of witnesses makes the case more interesting, but quantity alone cannot solve the central problems. Were witnesses independently describing the same creature? Had they previously heard the stories? Had earlier expeditions already introduced sauropod illustrations into the communities?

These questions matter because testimony can spread culturally. Once everyone knows that outsiders are searching for a long-necked dinosaur, later accounts are no longer completely independent.

The decisive evidence—a carcass, bone, DNA sample or unmistakable photograph—remains absent.

The Japanese Expedition Photograph

One of the strangest modern prehistoric-animal cases occurred far from any jungle. In 1977, the Japanese fishing vessel *Zuiyō Maru* hauled a decomposing carcass from waters near New Zealand. Photographs showed a bizarre body with what appeared to be a long neck, relatively small head and large flippers.

The carcass immediately generated speculation that fishermen had discovered a surviving plesiosaur.

This case is particularly important because something physical unquestionably existed. The fishermen were not simply reporting a distant shape. They photographed and measured a real carcass and retained tissue samples before disposing of the decomposing body.

Subsequent analysis strongly indicated that the carcass was a decomposed basking shark. When large sharks decay, their jaws and gill structures can disappear, producing a surprisingly long-necked, small-headed appearance.

The case demonstrates how a completely genuine encounter can generate a completely incorrect prehistoric interpretation.

No fraud is required.

The Ropen Witnesses

Papua New Guinea has generated repeated modern reports of large flying creatures interpreted by some cryptozoologists as pterosaurs. Witnesses describe enormous wings, strange silhouettes and occasionally long tails.

Some reports come from local residents familiar with ordinary regional wildlife, which proponents consider significant.

Yet estimating the size of a flying object is extremely difficult without a reference point. Large fruit bats, birds and other animals can appear enormous against an empty sky, especially at twilight.

Reports involving luminous flying creatures create additional complications because known pterosaurs provide no evidence of bioluminescence.

Again, there may be sincere witnesses without a prehistoric animal.

The Lake Monsters

Modern encounters with supposed prehistoric survivors are especially common around lakes. Scotland's Loch Ness provides the most famous example, although plesiosaurs were marine reptiles rather than dinosaurs.

Since the twentieth century, witnesses have reported long necks, humps, wakes and large unidentified shapes moving through the water.

Some photographs appeared extraordinary before being challenged or exposed as hoaxes. Others remain ambiguous because they show objects too distant to identify.

Lake monsters demonstrate one of the greatest weaknesses of eyewitness observation: **water hides anatomy**.

A witness sees a head in one location and a hump several meters behind it. The brain connects both objects into one enormous animal. Several swimming animals can resemble one serpentine creature. Birds, seals, large fish, floating logs and boat wakes can create surprisingly convincing monsters.

An observation can be honest and still be wrong.

Soldiers and Wartime Encounters

Stories involving soldiers possess particular psychological power because military witnesses are often assumed to be trained observers.

Reports occasionally circulate about soldiers encountering enormous reptiles or unusual flying creatures during operations in remote regions. Similar claims have emerged from wartime Africa, Southeast Asia and the Pacific.

But "soldier" does not automatically mean "expert zoologist."

Military personnel are trained for specific tasks. A pilot may be highly skilled at identifying aircraft while having no special ability to distinguish an unfamiliar bird at an uncertain distance. A soldier operating at night under stress may be less reliable when estimating the dimensions of an unknown animal.

Furthermore, many spectacular military cryptid stories emerge decades after the alleged events, making verification extremely difficult.

The delay between encounter and documentation is crucial.

Pilots and Objects in the Sky

Pilots occasionally report enormous birds or unidentified flying creatures. Their testimony deserves attention because pilots routinely estimate distance and movement.

Yet aerial perception also has limitations. Without a reliable reference point, even experienced observers can misjudge size and distance.

A nearby bird may appear like a much larger distant animal. Clouds and changing illumination alter apparent shapes. An observation lasting several seconds provides little opportunity for anatomical analysis.

A pilot reporting an unknown flying creature has established that the pilot could not identify what was seen—not that a pterosaur was present.

Fishermen and Sea Monsters

Fishermen probably encounter unusual animals more frequently than almost any other occupational group. They spend enormous amounts of time on water and occasionally haul creatures from depths rarely observed directly.

Stories include enormous serpentine animals, long-necked creatures and unfamiliar carcasses.

The oceans unquestionably contain species still unknown to science, particularly in deep water. This makes marine cryptozoology different from claims involving giant terrestrial animals living near human settlements.

But decomposition produces spectacular errors. Sharks and whales can become almost unrecognizable after death. Tissue disappears, skeletons separate and remaining structures create apparently impossible anatomy.

The *Zuiyō Maru* case provides an excellent warning: a "prehistoric monster" can be real, photographed and physically sampled—and still turn out to be a known animal.

Hunters in Remote Forests

Hunters are frequently presented as particularly credible witnesses because they are familiar with local wildlife. This can indeed make their observations more valuable.

If someone who has spent forty years hunting in a particular forest reports an animal unlike anything previously encountered, the testimony deserves investigation.

But familiarity does not eliminate perception errors.

A diseased animal may look radically different from a healthy one. Partial observation can hide identifying characteristics. An introduced species may be unfamiliar locally despite being known elsewhere.

The strongest hunter report would involve tracks, hair, scales, dung or biological material collected immediately after the encounter.

Most alleged dinosaur encounters provide none.

When the Story Changes

One of the most revealing investigative techniques is comparing the earliest account with later versions.

Cryptid stories often grow.

The original witness reports "a large dark animal."

Years later it becomes "a gigantic reptile."

Eventually it becomes "a fifteen-meter dinosaur."

A vague protrusion becomes a horn. A ridge becomes plates. An estimated size becomes a precise measurement.

This does not necessarily mean the original witness deliberately lied. Human memory is reconstructive. Every time we remember an event, later information can influence the reconstruction.

Repeated interviews can actually make the problem worse if investigators ask leading questions.

Independent Witnesses

Multiple witnesses can strengthen a case—but only if they are genuinely independent.

Five people discussing a strange object before being interviewed do not provide five fully independent descriptions. Their memories can influence one another.

The ideal case would involve witnesses observing the creature from different locations, recording their descriptions separately before communicating and producing consistent anatomical details.

Even stronger would be independent cameras documenting the same animal.

Very few living-dinosaur reports approach this standard.

The Camera Problem

The twenty-first century has created perhaps the greatest challenge for prehistoric-survivor claims.

Billions of people carry cameras.

Remote researchers use camera traps.

Tourists carry high-resolution phones.

Drones explore inaccessible terrain.

Satellites observe enormous areas.

Yet supposed living dinosaurs remain represented primarily by stories, distant shapes and ambiguous images.

This does not prove that every unknown animal has been discovered. Small, rare and secretive species can remain undocumented.

But for enormous animals supposedly encountered repeatedly, the absence of clear imagery becomes increasingly significant.

Sincere Does Not Mean Correct

The most important lesson from modern encounters is that testimony should not be divided simply into "truth" and "lie."

A third category is enormous:

sincere misidentification.

A witness can be intelligent, experienced and completely honest while incorrectly interpreting what was seen.

Human perception is not a camera. Our brains rapidly construct complete objects from incomplete information. Darkness, distance, fear, expectation and unfamiliar environments magnify errors.

This explains why eyewitness testimony can be fascinating while remaining scientifically insufficient.

The Encounter We Are Still Waiting For

Among thousands of modern stories, no encounter has yet produced universally convincing evidence of a living non-avian dinosaur or pterosaur. Some famous cases collapse under historical investigation. Others probably involve known animals. A smaller number remain genuinely unidentified because available information is insufficient.

And "unidentified" is an important word.

It means **we do not know**.

It does not mean dinosaur.

A truly persuasive modern encounter would need several elements together: independent witnesses, immediate documentation, clear photographs or video, tracks or biological material and subsequent investigation by researchers able to verify the evidence.

Such an encounter could happen tomorrow if prehistoric survivors really exist.

Until then, modern reports occupy an uncomfortable territory between zoology and human perception. They cannot prove that dinosaurs survived, but neither should every unexplained observation automatically be transformed into a joke or dismissed as deliberate fabrication.

Some witnesses almost certainly saw something.

The mystery is not whether they had an experience.

The mystery is **what was actually standing in the forest, moving through the water or crossing the sky when they did.**

Chapter 10 — Photographs, Videos, Footprints and Alleged Evidence

Stories can survive for centuries, but animals leave physical traces. If recognizable non-avian dinosaurs survived into modern times, somewhere there should be more than eyewitness testimony: photographs, tracks, bones, teeth, eggs, dung, skin, DNA or a carcass. This is the central weakness of the living-dinosaur hypothesis. The twentieth and twenty-first centuries have produced an enormous quantity of alleged evidence, yet no specimen has survived independent scientific examination and been confirmed as a modern non-avian dinosaur.

That does not mean every mysterious photograph or footprint is fraudulent. Some remain unidentified simply because the evidence is too poor to determine what produced them. The important distinction is between **unexplained evidence and evidence of dinosaurs**.

The Photograph Everyone Wants

Photography initially seemed capable of ending cryptozoological debates. If a strange animal existed, someone only needed to photograph it. Instead, photography created an entirely new category of mystery: the ambiguous image.

A distant object photographed through vegetation may contain only a few useful pixels. Water removes most of an animal's visible anatomy. Motion creates blur. Shadows manufacture limbs and heads that do not exist. Once viewers are told that an image supposedly shows a dinosaur, the brain begins searching for dinosaur-like features.

This psychological effect is powerful. A branch becomes a neck, a rock becomes a head and a wake becomes a long body.

A scientifically useful photograph would need much more: identifiable anatomical features, reliable scale, location data, preferably the original image file and ideally multiple photographs showing movement from different angles.

Most famous "living dinosaur" photographs fail these tests.

The Kasai Rex Images

Photographs associated with the alleged Kasai rex encounter in the Belgian Congo are among the most dramatic supposed pieces of dinosaur evidence. Circulating versions appear to show an enormous reptilian predator near a dead rhinoceros.

If authentic, such an image would be astonishing.

Unfortunately, the photographic evidence is highly problematic. Different images have circulated with the story, their provenance is uncertain, and the animals shown do not consistently correspond with descriptions attached to the alleged encounter. The pictures have been widely regarded as manipulated or otherwise unreliable.

The case demonstrates a crucial rule: an extraordinary photograph without a trustworthy chain of origin is weak evidence. We need to know who took it, when, where, with what equipment, where the original negative or file is, and whether the image existed before the story became popular.

Without provenance, an impressive image may be little more than an illustration attached to a legend.

Loch Ness and the Surgeon's Photograph

The famous 1934 "Surgeon's Photograph" appeared to show a long neck and small head emerging from Loch Ness. For decades it became one of the most recognizable monster photographs ever produced and strengthened the popular idea that a plesiosaur-like animal might inhabit the Scottish lake.

The photograph was eventually exposed as part of a hoax involving a small model.

The importance of the case extends far beyond Loch Ness. A single photograph influenced public imagination for decades because viewers saw what they expected to see: a prehistoric animal.

It demonstrates why photographic evidence must be authenticated independently rather than judged solely by appearance.

Videos and the Digital Age

Video should provide stronger evidence because movement is harder to misinterpret than a single frozen image. Yet alleged dinosaur videos usually present the same problems: distant objects, poor resolution, brief duration or uncertain origins.

The internet has made matters worse.

Computer-generated imagery can create convincing dinosaurs. Film footage can be edited. Frames can be removed from fictional productions and reposted without context. More recently, generative artificial intelligence has made fabricated imagery easier to produce.

Consequently, a spectacular video appearing online without provenance now has less evidential value than it might have possessed several decades ago.

The best modern evidence would require original files containing metadata, multiple independent recordings and a location where investigators could immediately search for physical traces.

Footprints: Evidence an Animal Was There

Tracks have an important advantage over eyewitness reports: they remain after the animal leaves.

A genuine modern dinosaur footprint could potentially reveal foot anatomy, number of toes, stride length, estimated body size and method of locomotion. A sequence of footprints would be substantially more valuable than one isolated impression.

But footprints are easily misinterpreted.

Mud changes shape as it dries. Rain enlarges impressions. Tracks made by known animals can overlap. Erosion can create footprint-like depressions. Human beings can manufacture tracks deliberately.

This is why a single enormous three-toed impression proves almost nothing.

A convincing trackway would require numerous consecutive prints showing consistent anatomy and natural variation. Researchers would photograph, measure and scan the tracks, examine sediment displacement and search the surrounding area for dung, feeding traces or additional tracks.

No modern trackway has been scientifically verified as belonging to a surviving non-avian dinosaur.

The Problem of Old Dinosaur Footprints

Another source of confusion is genuine fossil dinosaur tracks.

People around the world have discovered dinosaur footprints preserved in stone, sometimes long before anyone understood what dinosaurs were. Such tracks may have contributed to traditions concerning giant birds, dragons or monsters.

But a fossilized dinosaur footprint is evidence that a dinosaur walked there millions of years ago, not recently.

Determining the age of the material surrounding a footprint is therefore essential. A spectacular dinosaur-shaped track discovered in Cretaceous rock tells us nothing about modern survival.

Bones Above the Extinction Boundary

Bones seem more decisive, but even they can mislead.

As discussed earlier, dinosaur fossils have occasionally been reported from sediments apparently younger than the conventional extinction boundary. Such discoveries can generate claims that dinosaurs survived into the Paleocene.

The major alternative explanation is **reworking**. An old fossil can erode from Cretaceous sediment, move through water or geological disturbance and become redeposited inside younger material.

A dinosaur bone found in younger sediment is therefore not automatically a younger dinosaur.

Strong evidence of post-Cretaceous survival would require multiple associated remains securely embedded within younger geological contexts, ideally supported by additional evidence such as tracks, eggs or nests.

No generally accepted record demonstrates non-avian dinosaur populations continuing through the Cenozoic toward modern times.

Where Are the Recent Bones?

The modern survival hypothesis faces an even more direct problem.

Dinosaurs would die.

A breeding population surviving today would have produced countless dead individuals over generations. Even if tropical environments destroyed most remains quickly, some bones and teeth should survive long enough to be discovered.

Large dinosaurs would be particularly difficult to erase completely. A sauropod skeleton contains enormous bones. Predatory dinosaurs continually replaced teeth, creating additional material that could potentially survive.

Finding one fresh dinosaur tooth embedded in a recently killed animal would be vastly more significant than hundreds of eyewitness stories.

None has been authenticated.

Skin, Scales and Tissue

Claims occasionally appear concerning mysterious pieces of skin, scales or tissue supposedly belonging to unknown creatures.

Modern genetics makes these claims testable.

A sample does not need to look impressive. A tiny fragment can contain enormous biological information. DNA sequencing may determine whether material belongs to a known mammal, reptile, fish, bird or another organism.

This has transformed cryptozoological investigation.

Hair attributed to mysterious animals can be tested. Tissue from unidentified carcasses can be analyzed. Dung can reveal the animal that produced it.

A genuine dinosaur-lineage sample would produce extraordinary genetic results, although interpreting DNA from a lineage separated from known living groups by immense evolutionary distances would be complicated.

No verified modern sample has demonstrated a non-avian dinosaur.

Environmental DNA Changes the Search

Environmental DNA, or eDNA, may eventually become one of the most powerful tools for investigating alleged hidden animals.

Living organisms constantly release biological material into their surroundings through skin cells, mucus, waste, blood and other sources. Researchers can collect water or soil and analyze the DNA contained within it.

This means scientists do not necessarily need to see an animal to detect its presence.

If a large mysterious creature repeatedly inhabited a particular African lake, researchers could collect water samples from multiple locations and search for unusual vertebrate DNA. Repeated detection from independently collected samples would provide a powerful reason for further investigation.

Environmental DNA has limitations. DNA degrades, contamination occurs, databases are incomplete and unknown sequences can be difficult to classify.

But the technology dramatically reduces the ability of large animals to remain biologically invisible.

Camera Traps Never Sleep

Camera traps have revolutionized wildlife research. They can remain in forests for months, automatically photographing animals triggered by movement or heat.

Extremely elusive species are routinely documented this way.

A human explorer might walk through a forest for two weeks and see nothing. Fifty camera traps operating for six months provide a fundamentally different level of observation.

This creates another challenge for living-dinosaur claims. If investigators can identify a specific region where a large unknown animal is repeatedly reported, systematic camera-trap networks should eventually photograph it.

The fact that compelling dinosaur imagery has not emerged does not prove impossibility, particularly across enormous poorly surveyed regions, but every technological improvement reduces the remaining space for the hypothesis.

Drones Above the Wilderness

Drones add another layer. They can cross swamps inaccessible to people, follow rivers, inspect forest clearings and carry thermal cameras.

Dense rainforest canopy still presents a formidable barrier. A drone cannot simply see through several layers of vegetation. But large animals must occasionally cross rivers, clearings or open ground.

A population also occupies territory continuously. Searching once and finding nothing means little. Repeated systematic surveillance is much more powerful.

Future autonomous drones may make persistent wildlife monitoring increasingly inexpensive, further reducing biological blind spots.

What About Satellites?

Satellites have mapped almost the entire surface of Earth, but claims that satellites should be able to "see every dinosaur" misunderstand their capabilities.

A satellite image of a rainforest primarily records the canopy. An animal underneath may be completely invisible. Resolution, cloud cover, revisit frequency and vegetation all impose limitations.

Large animals in open landscapes are easier to detect, and modern remote sensing increasingly contributes to wildlife studies.

Therefore satellites make hidden populations less plausible in some environments, but they do not eliminate the possibility of unknown animals beneath dense forest.

The Smartphone Paradox

Perhaps the strongest cultural argument is simple: billions of humans now carry cameras.

Before 2000, a witness encountering something extraordinary could reasonably explain, "I didn't have a camera."

Today that explanation is weaker.

But unexpected encounters happen quickly. A frightened person may not reach for a phone. A creature may disappear within seconds. Night photography remains difficult. Remote communities may have limited technology.

So the absence of smartphone photographs cannot prove nonexistence.

Nevertheless, after thousands of alleged encounters across decades, we might reasonably expect at least a few clear recordings if large prehistoric animals regularly interact with humans.

The continued absence becomes statistically and scientifically significant.

What Would Count as Proof?

No single blurry photograph should rewrite paleontology.

A convincing case would probably emerge through several independent forms of evidence simultaneously: clear photographs, continuous video, footprints, environmental DNA, biological samples and repeated observations by independent researchers.

The strongest evidence would be unmistakable—a fresh carcass or living specimen.

Scientists could perform CT scans, anatomical examinations, radiometric investigation of surrounding material where relevant, genetic sequencing and comparisons with fossil dinosaurs and living animals.

The question could move from cryptozoology to zoology almost overnight.

Impossible or Extremely Unlikely?

Does the absence of physical evidence prove that non-avian dinosaurs cannot survive today?

Not absolutely.

Science rarely proves biological nonexistence everywhere on Earth. There are environments that remain difficult to survey, and unknown species undoubtedly await discovery.

But probability matters.

A small, rare animal living beneath dense tropical vegetation could remain undiscovered much more easily than a herd of twenty-ton herbivores. An unknown reptile is considerably easier to accommodate than a dinosaur lineage requiring a 66-million-year undocumented history.

Every year without physical evidence makes the most spectacular versions of the survival hypothesis harder to sustain.

The question has therefore changed.

A century ago explorers could reasonably say that enormous regions had barely been scientifically surveyed.

Today we possess camera traps, drones, genetic sequencing, environmental DNA, thermal imaging and global communication.

The hiding places have not disappeared, but they are shrinking.

If dinosaurs survived, they do not merely need to have escaped extinction.

They must also have escaped **fossilization, hunters, explorers, scientists, cameras, roads, aircraft, satellites, genetic surveys and billions of human observers.**

That is not logically impossible.

But after 66 million years, the absence of convincing physical evidence makes it an increasingly extraordinary possibility—and one that only equally extraordinary evidence could now establish.

Chapter 11 — The Scientific Case Against Living Dinosaurs

After examining legends, eyewitness accounts, mysterious photographs and unexplored regions, the investigation must confront the strongest objection of all: if non-avian dinosaurs survived into the modern world, **where is the biological evidence?** Science does not reject living dinosaurs merely because textbooks say they disappeared 66 million years ago. The hypothesis faces a collection of independent problems involving fossils, genetics, ecology, reproduction, population size and modern surveillance. None alone makes survival logically impossible, but together they create a formidable case.

The Missing 66 Million Years

The first problem is the fossil record. Non-avian dinosaurs are abundant in Mesozoic rocks, and then they disappear around the Cretaceous-Paleogene boundary. If even one lineage continued through the Cenozoic, paleontologists might expect to find traces in Paleocene, Eocene, Oligocene, Miocene, Pliocene or Pleistocene deposits.

The fossil record is incomplete, so a species can survive beyond its youngest known fossil. This is the Signor-Lipps problem discussed earlier. But incompleteness becomes progressively less persuasive as the proposed gap expands. Missing a population for several thousand years is one thing; proposing approximately 66 million years of survival without a convincing sequence of younger fossils is something entirely different.

A modern dinosaur would therefore require an enormous **ghost lineage**: generation after generation existing while leaving no generally accepted fossil trail toward the present.

One Dinosaur Is Not Enough

Living-dinosaur stories often focus on a single creature emerging from a river or crossing a road, but species do not survive geological ages as individuals. A reproductive population would be required continuously.

Minimum viable population depends on the species, environment and reproductive biology, so there is no magical number such as fifty or five hundred that applies to every animal. Nevertheless, extremely small populations are vulnerable to demographic accidents, disease, environmental catastrophes and genetic deterioration.

Imagine only twenty dinosaurs surviving in an isolated valley. A flood kills six. Several remaining animals are too old to reproduce. The next generation contains an unfavorable sex ratio. A disease then kills half the juveniles. Extinction can occur even when food is abundant.

Maintaining a lineage for millions of years would probably require larger populations and, ideally, movement between populations. But larger populations become much harder to hide.

This creates the living-dinosaur paradox: **enough animals must exist to prevent extinction, but few enough must exist to avoid detection.**

The Genetic Problem

Small isolated populations lose genetic diversity. Inbreeding increases the probability that harmful recessive variants will become expressed, while genetic drift can remove useful variation simply by chance. Over short periods a population may survive severe bottlenecks, as some living species demonstrate, but maintaining an isolated lineage through immense spans of geological time would require extraordinary persistence.

There is another consequence often overlooked by popular living-dinosaur stories: a population surviving 66 million years would evolve. Mutation, selection and genetic drift would continue. A modern descendant would not be genetically identical to its Cretaceous ancestors.

This does not mean it must transform into a completely different-looking animal. Evolution can preserve recognizable body plans for very long periods. A hypothetical surviving sauropod lineage might remain visibly sauropod-like. But describing it as an unchanged Cretaceous species would require evidence far beyond a superficial resemblance.

Giants Need Food

Ecology creates an even greater obstacle for large dinosaurs. A multi-ton herbivore needs a productive habitat containing enormous quantities of suitable vegetation. Hundreds of such animals would consume substantial biomass and alter the landscape through feeding and movement.

Large predators create another problem. A giant carnivore cannot exist without prey. A population of predators would require a much larger prey population beneath it in the food web. If the predator supposedly survives in an isolated ecosystem, an entire ecological system must support it.

A hidden animal weighing twenty kilograms is therefore fundamentally different from one weighing twenty tonnes. The larger the species, the more ecological disturbance it creates.

A creature may hide visually while remaining ecologically obvious.

Breeding Leaves Evidence

Dinosaurs reproduced. Most known dinosaurs laid eggs, constructed or used nesting sites in various ways, and produced juveniles. A surviving population would therefore leave more than adult footprints.

There should be eggshells, nests, juvenile tracks and failed eggs. Breeding grounds might be used repeatedly, increasing the chances of discovery.

Even if a hypothetical modern lineage evolved different reproductive behavior, successful reproduction would still require regular biological activity somewhere.

A population cannot remain permanently concealed as a collection of mysterious adults appearing individually every few decades.

There must be young animals.

Dinosaurs Would Die

Perhaps the strongest simple argument is mortality.

Every dinosaur eventually dies.

If a population contained hundreds of animals, generations would continuously produce carcasses. Tropical decomposition can be extremely rapid, and acidic soils can destroy bones, so we should not imagine pristine skeletons waiting on every forest floor. Scavengers disperse remains, rivers transport them and vegetation conceals them.

But millions of years provide enormous numbers of opportunities for preservation and discovery.

Large dinosaurs possessed large bones. Predatory dinosaurs replaced teeth throughout their lives. Animals crossing muddy ground produced tracks. Injured individuals left blood and tissue. Dead animals entered rivers. Floods buried remains.

A long-lived population should leave something.

DNA Makes Hiding Harder

Modern molecular biology has changed the search for unknown animals. Researchers no longer need to capture a creature to detect it.

Hair, scales, blood, saliva, dung and tissue can contain DNA. Environmental DNA goes further by recovering genetic material organisms leave in water, soil and other environments.

Suppose Mokele-mbembe regularly inhabits one particular lake. Researchers could repeatedly sample the water. Known fish, mammals, reptiles and microorganisms would produce recognizable genetic signatures. An unusual large vertebrate repeatedly generating unknown sequences would immediately attract attention.

Environmental DNA is not perfect. Material degrades, samples become contaminated and reference databases are incomplete. But the technology makes complete biological invisibility increasingly difficult.

Humans Have Expanded Almost Everywhere

A dinosaur population surviving until 10,000 years ago would face one problem. Surviving into the twenty-first century creates another: us.

Humans have entered forests, drained wetlands, constructed roads, mined mountains, dammed rivers and settled islands. Hunters penetrate remote landscapes. Logging operations reach forests once considered inaccessible. Aircraft cross wilderness areas. Scientists tag animals, install camera traps and collect biological samples.

This expansion has been disastrous for many known species, but it also means large animals increasingly encounter humans.

A population of giant dinosaurs would need not only to survive natural environmental changes but also to avoid centuries of expanding human activity.

Wildlife Surveys Never Stop

Modern zoology possesses tools earlier explorers could hardly imagine: automatic cameras, acoustic recorders, thermal imaging, drones, satellite tracking, environmental DNA and large-scale biodiversity surveys.

Camera traps are especially damaging to the argument that large terrestrial animals can easily remain invisible. They routinely photograph shy species that researchers almost never encounter directly.

Unknown species certainly remain undiscovered, but modern survey technology changes the likely characteristics of those species. The easiest animals to overlook are generally small, rare, nocturnal, aquatic, subterranean, canopy-dwelling or restricted to extremely inaccessible habitats.

A herd of giant terrestrial herbivores represents almost the opposite profile.

But Science Still Finds New Animals

Here the counterargument becomes important. Declaring that every large animal has already been discovered would be incorrect.

Zoological discoveries continue. Sometimes scientists formally describe animals known locally for generations but previously overlooked by international science. Remote oceans, forests and caves remain incompletely catalogued. Even relatively conspicuous species can escape formal recognition when populations are small or geographically restricted.

This means "we haven't found it" cannot automatically become "it cannot exist."

History provides one particularly famous example.

The Coelacanth

Coelacanths are frequently invoked whenever supposedly extinct creatures are discussed. The group was known from fossils, and scientists once believed its lineage had disappeared around the end of the Cretaceous. Then, in 1938, a living coelacanth was famously recognized from a specimen caught off southern Africa.

It was an extraordinary discovery.

But the analogy with dinosaurs is often exaggerated. The modern coelacanth is not simply an individual Cretaceous fossil species that remained perfectly unchanged. It belongs to a lineage with a long evolutionary history, and its deep-water marine environment is fundamentally different from the habitats proposed for giant terrestrial dinosaurs.

The coelacanth proves an important but narrower point: **absence from a known recent fossil record does not guarantee that an entire lineage is extinct.**

It does not prove that any extinct lineage can survive anywhere for any length of time.

How Big Could an Unknown Animal Be?

This is ultimately the more useful question.

Could an unknown insect exist? Certainly.

An unknown frog or lizard? Easily.

A previously undescribed fish? Very plausibly.

A medium-sized mammal in remote rainforest? Possible.

A large aquatic animal in poorly surveyed waters? More difficult, but conceivable.

A five-ton terrestrial animal requiring a breeding population, enormous food resources and substantial territory? Much harder.

A twenty-ton sauropod-like animal surviving in numbers sufficient to maintain genetic viability while leaving no verified bones, eggs, dung, tracks, photographs or DNA? Extraordinary.

Size is not the only factor. Habitat, behavior, population density and metabolism matter. A secretive nocturnal predator occupying dense forest could be harder to detect than a smaller herd animal crossing open landscapes.

The real variable is **ecological conspicuousness.**

The Strongest Case Against Survival

Science cannot demonstrate that no non-avian dinosaur exists in every unexplored corner of Earth. Absolute biological nonexistence is extraordinarily difficult to prove. What science can do is compare a hypothesis with the evidence that should exist if the hypothesis were true.

For large living dinosaurs, we would expect populations. Populations require food and breeding. Breeding produces juveniles and eggs. Animals create tracks and waste. Individuals die and produce carcasses and bones. Populations leave DNA. Large creatures alter ecosystems. Humans eventually encounter them, and modern technology increases the probability that those encounters will be recorded.

Yet after decades of expeditions and centuries of stories, none of these evidence streams has produced a scientifically verified modern non-avian dinosaur.

That does not make survival philosophically impossible.

It makes it **extraordinarily unlikely**, particularly for large species.

The strongest remaining possibility would involve something small, rare, geographically restricted and exceptionally difficult to detect. Ironically, such an animal would look very different from the giant sauropods, horned dinosaurs and enormous predators dominating living-dinosaur legends.

The mystery therefore survives, but in a narrower form.

The question is no longer simply, "**Can unknown animals still exist?**"

They unquestionably can.

The scientifically meaningful question is:

How large, numerous and ecologically conspicuous could an animal be before its continued absence from fossils, DNA surveys, cameras and physical collections becomes stronger evidence than the stories claiming it survives?

Chapter 12 — Somewhere, Could They Still Be Alive?

After ancient chronicles, dragon traditions, African cryptids, Amazonian monsters, alleged pterosaurs, modern encounters, photographs, footprints and biological arguments, we return to the question that began this investigation: **could recognizable non-avian dinosaurs still be alive somewhere on Earth?**

The scientifically established answer is straightforward. There is currently no verified evidence demonstrating that non-avian dinosaurs survived into the modern era. No living specimen has been scientifically documented. No recent carcass has been authenticated. No modern dinosaur DNA has been confirmed. No population has been photographed repeatedly and clearly. The accepted fossil record places the disappearance of non-avian dinosaurs at approximately 66 million years ago.

But there is an important difference between saying "**there is no evidence that this happened**" and saying "**this could not possibly have happened under any circumstances.**"

Science works with evidence and probabilities, not with the requirement that every imaginable scenario be declared absolutely impossible. The living-dinosaur hypothesis therefore becomes more useful when divided into different levels of plausibility rather than treated as one enormous yes-or-no question.

Level One: Individuals Surviving the Catastrophe

The easiest scenario to accept is that some individual non-avian dinosaurs survived the immediate catastrophe at the end of the Cretaceous.

Mass extinction does not necessarily mean every member of an affected group dies simultaneously. The asteroid impact and its environmental consequences unfolded through a sequence of disasters: impact effects, fires, darkness, climatic disruption, ecological collapse and food shortages.

Individual dinosaurs could have survived the initial event only to die weeks, months or years later.

This would not challenge the accepted mass extinction. It would simply mean that extinction was a process rather than one instantaneous moment.

Of all the scenarios examined in this book, short-term survival is by far the least extraordinary.

Level Two: Populations Surviving Beyond the Boundary

More interesting is the possibility that small breeding populations persisted beyond the conventional extinction boundary.

Perhaps an isolated ecosystem escaped the worst consequences. Perhaps small omnivorous dinosaurs found enough food to reproduce for several generations. Fossilization is rare, so a geographically restricted population might theoretically disappear without leaving an easily recognizable record.

Such survival could last decades, centuries or perhaps longer.

The Signor-Lipps effect reminds us that the youngest fossil we have discovered does not necessarily represent the final individual that lived.

But as the proposed survival interval grows, the evidential problem becomes progressively larger.

Thousands of years might be difficult to detect.

Millions of years are something else entirely.

Level Three: Survival for Millions of Years

Suppose a dinosaur lineage survived one million years beyond the K–Pg extinction.

It would experience thousands of generations. Individuals would reproduce and die. Populations might expand or contract. Rivers would occasionally bury carcasses. Teeth, eggs and bones would enter sediments.

Extend the survival period to ten million years and the missing evidence becomes harder to explain.

Extend it to thirty million.

Fifty million.

Sixty-six million.

Eventually the hypothesis requires not merely an incomplete fossil record but an enormous undocumented evolutionary history.

This is possible to imagine, but increasingly difficult to support scientifically.

A species surviving for such an immense period would also evolve. Therefore, if descendants of a non-avian dinosaur lineage existed today, we should not expect them to be perfect copies of a Cretaceous species.

They could remain recognizably dinosaur-like while accumulating evolutionary changes.

Level Four: A Small Dinosaur Surviving Today

Now consider size.

A surviving dinosaur weighing two kilograms presents a fundamentally different problem from one weighing twenty tonnes.

A small animal could hide beneath vegetation. It might be nocturnal. It could occupy caves, tree canopies, remote islands or dense tropical forest. Its carcass might disappear rapidly through decomposition and scavenging. Its tracks could easily be overlooked.

Thousands of small animals can inhabit an ecosystem without attracting much human attention.

If some unknown non-avian dinosaur lineage survived into the present, this is probably the least implausible modern scenario: **small, rare, secretive and geographically isolated**.

There is still no evidence that such an animal exists.

But biologically, it would be easier to conceal than the giants of cryptozoological tradition.

Level Five: The Hidden Giants

Mokele-mbembe represents the opposite extreme.

A multi-ton sauropod-like animal requires enormous resources. It creates pathways through vegetation, eats large quantities of plants, produces substantial dung and leaves large tracks. A viable population would require multiple breeding individuals.

Some would die.

Their bones would remain somewhere.

A population of giant predators would create similar problems while additionally requiring a substantial prey base.

This does not make concealment logically impossible in every conceivable circumstance. Dense rainforest and enormous wetlands can hide large animals remarkably well from direct human observation.

But hiding an individual for several days is not the same as hiding a species for millions of years.

The larger the proposed dinosaur becomes, the greater the evidence we should expect.

Geography Matters

Location changes probability dramatically.

A dinosaur living near Paris, New York, Athens or Tokyo would have virtually no realistic chance of remaining undocumented for generations. Roads, buildings, traffic, cameras and millions of people would make repeated encounters inevitable.

The Congo Basin presents a different environment.

So does Amazon.

New Guinea.

Remote Indonesian islands.

Certain enormous wetlands.

Isolated mountain forests.

These places remain biologically difficult to survey comprehensively even though their geography is mapped.

Satellites know where the forest is.

They do not necessarily know everything underneath it.

Nevertheless, indigenous communities often possess detailed knowledge of local wildlife. A giant animal might remain unknown to international science while being familiar locally, but it would be much harder for it to remain unknown to everyone.

This is why careful investigation of indigenous testimony matters—provided researchers avoid contaminating accounts by introducing dinosaur images or leading questions.

What Discovery Would Change Everything?

The debate could be transformed instantly.

Not by another story.

Not by another distant shape photographed through trees.

Not by another anonymous internet video.

Physical evidence is required.

A **fresh carcass** would be extraordinary. Scientists could examine anatomy, tissues and genetic material while establishing where and when the animal died.

Recent bones could also transform the debate if their age and geological context were demonstrated beyond reasonable doubt. A dinosaur bone from Cretaceous rock proves nothing about modern survival; a genuinely recent bone belonging to a non-avian dinosaur lineage would be revolutionary.

DNA would be even more powerful. Tissue, blood, scales, dung or other biological material could reveal evolutionary relationships.

Then there are **eggs**. A recently constructed nest containing embryos anatomically identifiable as non-avian dinosaurs would demonstrate not merely survival but active reproduction.

The Environmental DNA Revolution

Perhaps the decisive discovery will not begin with someone seeing a monster.

It may begin inside a laboratory.

Environmental DNA allows researchers to search for animals through genetic material released into water, soil and other environments. If an unknown large vertebrate repeatedly uses a particular river or lake, it should release biological traces.

One unusual DNA sequence would not be enough. Contamination, sequencing errors and incomplete databases must be considered.

But imagine researchers sampling a remote Congo lake.

An unknown vertebrate sequence appears.

Another team collects new samples independently.

The sequence appears again.

A third laboratory confirms it.

Samples from nearby rivers reveal the same genetic signature.

Camera traps are deployed.

Tracks are discovered.

Then a clear image appears.

Scientific revolutions often occur through this accumulation of independent evidence rather than one dramatic moment.

The Photograph That Would Matter

Photography could still provide the breakthrough, but standards have changed.

A single photograph is no longer enough.

Digital manipulation and artificial intelligence can manufacture convincing creatures that never existed. Future evidence would therefore require authentication.

Repeated high-resolution images from independent cameras would be far stronger. Continuous video showing natural movement would add evidence. Multiple observers photographing the same animal from different positions would strengthen the case dramatically.

Then researchers would need to find the animal itself.

An image should begin an investigation, not end it.

A Living Specimen

Ultimately, nothing would be more decisive than a living animal.

Imagine an expedition documenting a large unknown reptilian creature in an isolated forest. Zoologists observe it directly. Veterinarians obtain biological samples without harming it. Genetic laboratories analyze the DNA. Anatomists compare its skeleton with fossil groups.

If the results demonstrated that the animal belonged to a non-avian dinosaur lineage, biology and paleontology would change immediately.

Textbooks would be rewritten.

The K–Pg extinction would need reinterpretation.

Paleontologists would begin searching Cenozoic deposits for the missing evolutionary history.

Conservationists would face an urgent problem: a species surviving tens of millions of years could potentially be destroyed within decades after discovery.

The scientific question would no longer be whether dinosaurs survived.

It would become how they survived.

What If We Never Find Anything?

The opposite possibility must be taken equally seriously.

Perhaps every Mokele-mbembe story ultimately involves folklore, elephants, hippos, crocodiles, floating vegetation or misinterpretation.

Perhaps every ropen is a bat, bird, distant light or cultural tradition.

Perhaps giant Amazonian monsters are anacondas, caimans and stories enlarged through generations.

Perhaps medieval dragons were combinations of snakes, crocodiles, fossils, religious symbolism and imagination.

Perhaps there are no surviving non-avian dinosaurs anywhere.

This explanation currently requires fewer extraordinary assumptions and fits the available physical evidence better than modern dinosaur survival.

That is why living dinosaurs remain outside established zoology.

But Unknown Does Not Mean Empty

None of this makes exploration pointless.

Quite the opposite.

The search for legendary animals can lead researchers into poorly studied ecosystems, preserve local knowledge and encourage biological surveys. Even when the monster disappears under investigation, something scientifically valuable may remain.

The discovery does not need to be a dinosaur to matter.

An unknown reptile, mammal, fish or amphibian would be a genuine zoological discovery.

The history of science repeatedly demonstrates that nature is capable of surprising us.

But surprise should never become an excuse for abandoning standards of evidence.

The Last Blank Spaces

For most of human history, maps contained enormous unknown regions. Beyond familiar territory lay forests, mountains, islands and oceans where almost anything seemed possible.

Those blank spaces inspired dragons.

Sea serpents.

Lost civilizations.

Giant animals.

And eventually living dinosaurs.

Today the geographical blanks have largely disappeared. Satellites have photographed the continents. Aircraft cross remote regions. GPS can locate almost any expedition.

But biological blank spaces remain.

We still do not know every species inhabiting Earth's forests, oceans, caves and soils.

Somewhere in those remaining spaces, extraordinary animals almost certainly await discovery.

Whether any belong to dinosaur lineages thought extinct for 66 million years is another matter entirely.

There is currently no scientifically verified evidence that they do.

Perhaps that will never change.

Perhaps every expedition will return without the creature it sought.

Or perhaps, somewhere beneath the canopy of a forest that satellites can map but cannot fully penetrate, an animal survives that should have disappeared millions of years ago.

Until physical evidence appears, it remains speculation.

But the question that has followed humanity from ancient dragon legends into the age of DNA sequencing remains irresistible:

Did the dinosaurs disappear with the last blank spaces on our maps—or is there still somewhere on Earth where the age of dinosaurs never completely ended?

Also by Andrei Nebeleac

Did Dinosaurs Survive?
Can AI Beat the Casino?

Watch for more at https://www.nebeleac.com.

About the Author

Andrei Nebeleac is a Romanian author with a multidisciplinary background in literature, computer science, economics, and territorial management. Born in Sighetu Marmației, Romania, he comes from a literary and academic family and made his literary debut in 1997, publishing poetry in the Romanian literary magazine *Glasul Literar și Artistic.*

His writing explores a wide range of subjects, including history, science, unexplained phenomena, technology, economics, business, and the unusual questions that exist at the boundaries of established knowledge.

Combining analytical investigation with accessible storytelling, Nebeleac is particularly interested in subjects where documented facts, historical accounts, competing theories, and unresolved mysteries intersect.

Read more at https://www.nebeleac.com.

www.ingramcontent.com/pod-product-compliance
Lightning Source LLC
LaVergne TN
LVHW100147061126
851857LV00010B/273
* 9 7 9 8 2 3 5 6 8 6 9 7 7 *